Springer Theses

Recognizing Outstanding Ph.D. Research

For further volumes:
http://www.springer.com/series/8790

Aims and Scope

The series "Springer Theses" brings together a selection of the very best Ph.D. theses from around the world and across the physical sciences. Nominated and endorsed by two recognized specialists, each published volume has been selected for its scientific excellence and the high impact of its contents for the pertinent field of research. For greater accessibility to non-specialists, the published versions include an extended introduction, as well as a foreword by the student's supervisor explaining the special relevance of the work for the field. As a whole, the series will provide a valuable resource both for newcomers to the research fields described, and for other scientists seeking detailed background information on special questions. Finally, it provides an accredited documentation of the valuable contributions made by today's younger generation of scientists.

Theses are accepted into the series by invited nomination only and must fulfill all of the following criteria

- They must be written in good English.
- The topic should fall within the confines of Chemistry, Physics, Earth Sciences, Engineering and related interdisciplinary fields such as Materials, Nanoscience, Chemical Engineering, Complex Systems and Biophysics.
- The work reported in the thesis must represent a significant scientific advance.
- If the thesis includes previously published material, permission to reproduce this must be gained from the respective copyright holder.
- They must have been examined and passed during the 12 months prior to nomination.
- Each thesis should include a foreword by the supervisor outlining the significance of its content.
- The theses should have a clearly defined structure including an introduction accessible to scientists not expert in that particular field.

Aliaksei Charnukha

Charge Dynamics in 122 Iron-Based Superconductors

Doctoral Thesis accepted by
the University of Stuttgart, Germany

Author
Dr. Aliaksei Charnukha
Solid-State Spectroscopy
Max Planck Institute for Solid State Research
Stuttgart
Germany

Supervisor
Prof. Dr. Bernhard Keimer
Max Planck Institute for Solid State Research
Stuttgart
Germany

ISSN 2190-5053 ISSN 2190-5061 (electronic)
ISBN 978-3-319-37752-0 ISBN 978-3-319-01192-9 (eBook)
DOI 10.1007/978-3-319-01192-9
Springer Cham Heidelberg New York Dordrecht London

Softcover reprint of the hardcover 1st edition 2014

Printed on acid-free paper

Springer is part of Springer Science+Business Media (www.springer.com)

Supervisor's Foreword

The recent discovery of high-temperature superconductivity in iron-based superconductors has sparked a worldwide research effort to uncover the origin of this phenomenon. The goal of the present thesis was to obtain insight into electronic correlations and the mechanism of Cooper pair formation in iron arsenide superconductors with the so-called "122" structure from accurate optical spectroscopy data.

With this motivation, Aliaksei Charnukha has used spectroscopic ellipsometry over a wide range of photon energies in order to develop a comprehensive, highly accurate description of the optical conductivity of $Ba_{0.68}K_{0.32}Fe_2As_2$, a prototypical iron-based high-temperature superconductor with minimal chemical disorder. He analyzed the superconductivity-induced infrared anomalies he observed in terms of an Eliashberg model with a bosonic mode whose characteristics are compatible with a spin fluctuation mode independently observed in the same compound, and thus made a compelling case for a magnetically mediated Cooper pairing mechanism. He also showed that the analysis scheme he developed provides an excellent description of previously published data on other, chemically more disordered iron arsenide superconductors. Based on high-energy ellipsometric data with unprecedented accuracy, he went on to discover a highly unusual superconductivity-induced spectral anomaly at a photon energy of 2.5 eV, more than two orders of magnitude larger than the superconducting energy gap. While a quantitative description of this unexpected discovery has yet to be developed, Aliaksei provided an insightful interpretation based on the multiband nature of the superconducting state.

Aliaksei's thesis work comprises a second, equally comprehensive research project on the phase behavior and optical properties of $Rb_2Fe_4Se_5$, a compound that has attracted considerable attention because of the close competition between antiferromagnetic Mott-insulating and high-temperature superconducting phases. By again combining high-quality ellipsometric data with an incisive analysis, he showed that the surprising optical properties of this compound (which is optically transparent, but develops superconductivity at low temperatures) can be explained as a consequence of nanoscopic coexistence of insulating and superconducting phases. He subsequently used a combination of two complementary experimental methods—scanning near-field optical microscopy and low-energy muon spin rotation—to directly image the phase coexistence and quantitatively determine the phase composition as a function of depth below the surface. His data will have important implications for the interpretation of data from other experimental probes, especially those that are surface sensitive.

Stuttgart, June 2013 Prof. Bernhard Keimer

Acknowledgments

I would like to thank Prof. Keimer for giving me the opportunity to carry out my doctoral research in his group at the Max Planck Institute for Solid-State Research and Dr. Alexander Boris for introducing me to spectroscopic ellipsometry and supervising my experimental activities at the institute, as well as for their constant support and advice throughout three and a half fruitful years I had a chance to spend in this fertile scientific environment.

I am also grateful to the many great scientists at the institute with whom I had a chance and pleasure to discuss scientific issues and more at various times of my stay: Chengtian Lin, Reinhard Kremer, Jörg Smet, Markus Lippitz, Lilia Boeri, Luciano Ortenzi, Deniss Grjaznovs, Evgeny Blokhin, Eugene Kotomin, and many others. Even more so I am thankful to all previous and current members of our department, in particular Dmitro Inosov, Alexander Yaresko, Oleg Dolgov, George Jackeli, Giniyat Khaliullin, Santiago Blanco, Mathieu LeTacon, Darren Peets, Philippe Leininger, Martin Rahlenbeck, Paul Popovich, Heiko Uhlig, Benjamin Bruha and all others, who all no doubt contributed to the overall stimulating and healthy scientific working and personal atmosphere in the group and were often great travel mates at many conferences all over the globe.

My deep personal thanks go to all those whom I have had a chance to have as my office mates and with whom I continuously had a great time and who made the office my second home: Daniel Haug, Jitae Park, Alex Fraño, Michaela Souliou, and Friederike Wrobel. I would like to express my particular gratitude to Daniel Pröpper, with whom I spent countless hours doing experiments both in Germany and in the United States and who has proven to be a great colleague and friend, with very similar interests and habits.

I would also like to thank all those with whom I have worked at various external facilities: Michael Süpfle and Ives-Laurent Mathis at the ANKA synchrotron of Forschungszentrum Karlsrue; Andrei Sirenko, Taras Stanislavchuk, Eric Standard, and Larry Carr at the National Synchrotron Light Soruce of Brookhaven National Laboratory; Thomas Prokscha and Andreas Suter at Paul-Scherrer-Institut Villigen, Switzerland; Antonija Cvitkovic, Nenand Ocelic, Rainer Hillenbrand, and Fritz Keilmann at NeaSpec in Martinsried (Munich), Germany.

Last but not least I am grateful to my parents, brother, and my love Justine for their continuous support throughout my doctoral research.

Contents

Abbreviations

AFM	Atomic-force microscopy
ARPES	Angle-resolved photoemission spectroscopy
BCS	Bardeen, Cooper, Schrieffer
BKFA	$Ba_{0.68}K_{0.32}Fe_2As_2$
DOS	Density of states
INS	Inelastic neutron scattering
KFS	$K_2Fe_4Se_5$, $K_xFe_{2-y}Se_2$
KK	Kramers–Kronig
LDA	Local-density approximation
LE-μSR	Low-energy muon-spin rotation/relaxation
NMR	Nuclear magnetic resonance
OC	Optical contrast
QO	Quantum oscillations
RFS	$Rb_2Fe_4Se_5$, $Rb_xFe_{2-y}Se_2$
STM/STS	Scanning tunneling microscopy/spectroscopy
SW	Spectral weight
SDW	Spin-density wave
s-SNOM	(Apertureless) Scattering-type scanning near-field optical microscopy

Chapter 1
Introduction

If I have seen further it is by standing on the shoulders of giants.
—Isaac Newton, *Letter to Robert Hooke*

1.1 General Overview

Observation of the outside world is one of the most ancient human activities. All the civilizations, or at least those of which records remain, observed, analyzed, and inferred. Human intellect possesses the amazing power to observe the outside world and distill these observations in a purely abstract way into a set of logical inferences that capture the world's essence and explain its workings. Our ancestors, by a mere act of reasoning, perceived some of the fundamental properties of matter that it took us millennia to understand in detail but which hardly changed in essence from this primordial understanding. The most prominent of such examples are the 'classical elements' that many philosophies introduced at different stages in human history but which show a significant overlap among civilizations. One of the most structured formulations was made by the Greeks, who distilled the world's essence into a set of elements: Earth, Water, Air, Fire, and Aether. It is remarkable that this original classification remains largely valid today and any deviation from it has stimulated an explosion of research activities upon its discovery. Granted, our current scientific understanding of the world is rather complex, but its essence shows striking parallels with the classical elements of ancient Greece. Aether, an omnipresent entity from which everything is born to our world and to which it all eventually decays, has the fundamental properties of 'energy' as we know it today. All known elementary particles can be created from sheer energy, either spontaneously (such as the spontaneous creation of electron-positron pairs in black holes, which leads to their gradual evaporation [1]) or in collisions of various energetic particles, which lead to jets of newborn particles emanating in all directions. Every particle is believed to have an

A. Charnukha, *Charge Dynamics in 122 Iron-Based Superconductors*,
Springer Theses, DOI: 10.1007/978-3-319-01192-9_1,

antiparticle, many of which have been observed, with which it can annihilate so as to reduce itself back to energy. This concept of energy is valid at the very fundamental level of our understanding of the universe. Earth, Water, and Air remained the only known fundamental states of matter until the discovery of another one—plasma—by Sir William Crookes in 1879. Interestingly, and this only emphasizes the point being made, although Fire does not represent a state of matter, high enough temperature in a flame can give rise to ionization and thus plasma. Thus the ancient classical elements all have a place in modern science. The development of new tools and techniques, most notably cryogenics and high-energy colliders but also many others, has since lead to the discovery of many new states of matter. One of them, realized at the dawn of the twentieth century, is the subject of the present thesis. It was in 1911 that the production of extreme cryogenic temperatures by Heike Kamerlingh Onnes in Leiden, the Netherlands led to his discovery of the fifth state of matter in the history of humankind—the superconducting state.

It took a rather long time after the original discovery of superconductivity by Onnes in 1911 until theoretical understanding of this complex phenomenon started to take shape with the London equations of superconductivity put forward in 1935 and a more sophisticated macroscopic Ginzburg-Landau theory formulated in 1950 [2, 3]. Both approaches were essentially phenomenological in that they provided a good description of the macroscopic properties of superconductors known to date but irrespective of the microscopic properties of the host materials. It was not before 7 years later that the first coherent microscopic theory of this phenomenon was proposed by John Bardeen, Leon Neil Cooper, and John Robert Schrieffer (BCS) in 1957 [4–6]. Lev Petrovich Gor'kov showed 2 years later that all the phenomenological parameters of the Ginzburg-Landau theory can be derived consistently in the framework of the BCS theory in a certain limit [7]. The BCS theory provided such a satisfactory description of the fundamental properties of all superconducting materials at the time that superconductivity was considered a largely solved problem of contemporary physics. This *status quo* remained in the field for almost three decades until 1986, when Karl Müller and Johannes Bednorz discovered an entirely different family of superconductors based on copper-oxygen atomic planes [8]. Some of the compounds of this family were later found to exhibit transition temperatures as high as 135K, as shown in Fig. 1.1.

These new materials turned out to have very different properties from those of all previously known conventional superconductors. The latter mostly comprised elemental materials such as Hg, Pg, Nb etc. as well as simple binaries like V_3Si, Nb_3Sn, Nb_3Ge. All of these conventional materials are trivial solids below room temperature without any additional phases besides a superconducting and a normal paramagnetic state. Superconductivity in all of them is well described either by the BCS theory, with the superconducting state realized by pairs of electrons weakly attracted to each other via phonons, or lattice distortions, or, in the case of stronger attraction, by an extension of the BCS theory taking a full account of the retarded electron-phonon interaction — the so-called Migdal-Eliashberg theory of superconductivity [9]. The copper-based superconductors, on the other hand, immediately revealed close proximity of superconductivity to strong local antiferromag-

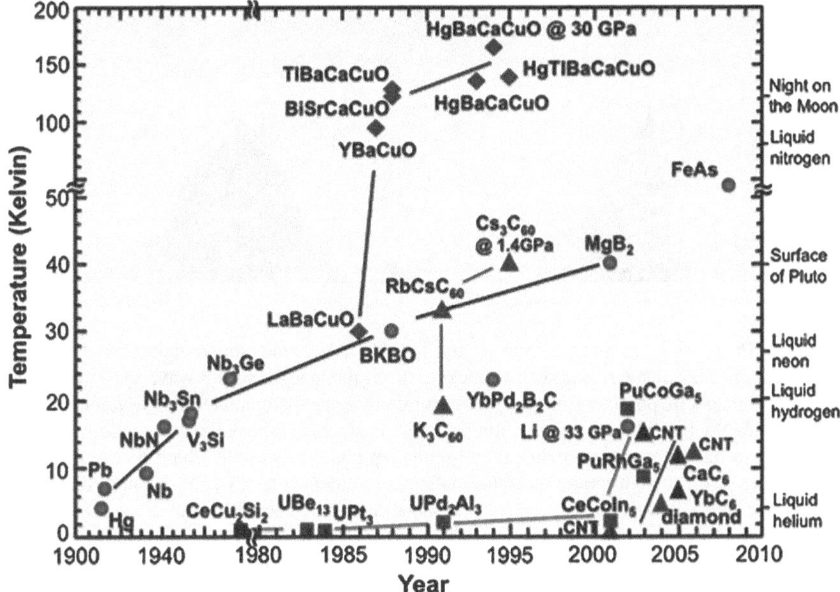

Fig. 1.1 Superconducting transition temperature of several representative materials at the moment of their discovery: conventional superconductors (*red circles* up to MgB_2), copper-based high-temperature superconductors (*blue diamonds*), heavy-fermion compounds (*red squares*), and the highest transition temperature in the recently discovered family of iron-based superconductors (*red circle*, FeAs). Adapted from Wikipedia (History of superconductivity). This image is in the public domain and free to reuse. Originally submitted by Department of Energy: Basic Energy Sciences

netism [10–12] and, in the following years, to other phases and instabilities [13–19], all of which show an intimate connection with superconductivity of the character of either competition or cooperation, giving rise to rather complex phase diagrams like that shown in Fig. 1.2a, which have seen many additions and corrections over the course of almost three decades since the original discovery of these materials. This richness of the phase diagram of the cuprates, with several coexisting phases present, significantly complicated data analysis for all experimental techniques. Only today, having improved the resolution of many different spectroscopic techniques such as angle-resolved photoemission spectroscopy (ARPES), resonant elastic/inelastic X-ray scattering (REXS/RIXS), etc., do we finally reach the required experimental capacity to disentangle and understand the multiple phases of these fascinating materials.

The complexity of the phase diagram notwithstanding, already early on it became clear that magnetism plays a special role in the superconductivity of the cuprates. Although proximity of superconductivity to some form of magnetism has been observed in various materials (heavy-fermion, organic and other superconductors), only in the cuprate superconductors does this proximity result in very high transition temperatures and seems to foster rather than destroy the superconducting state.

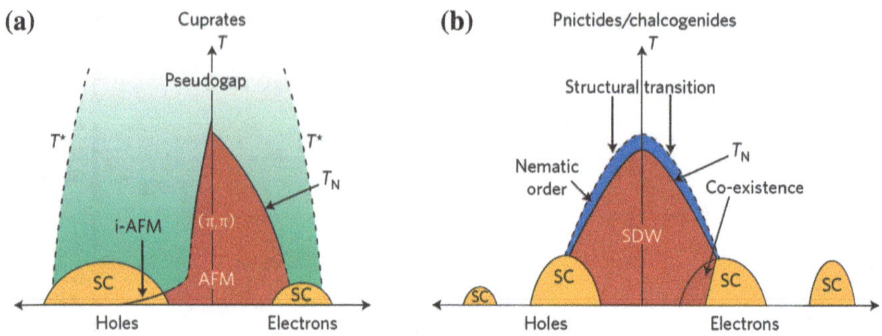

Fig. 1.2 **a** Phase diagram of the electron- and hole-doped cuprate superconductors. The parent compounds are Mott insulators and show local antiferromagnetism with a wave vector of (π, π). The antiferromagnetic phase region (*red*) extends into that of superconductivity (*yellow*) and in the overlap region the two either coexist or compete with each other. Above the Néel and/or superconducting transition temperature the so-called pseudogap phase is present, which breaks the tetragonal symmetry of the high-temperature normal state reducing it to C_2 [20]. **b** Phase diagram of the electron- and hole-doped iron-based superconductors. The general features are similar to those of the cuprate superconductors: superconductivity in the otherwise non-superconducting antiferromagnetic parent compounds can be induced by doping charge carriers of either sign, which results in a weakening of the itinerant antiferromagnetic phase (*red*) and appearance of a dome-shaped superconducting phase region (*yellow*) with coexistence/phase separation of the superconducting and antiferromagnetic phase at certain doping levels. Superconductivity has also been found to occur in the extremely overdoped regime either disconnected from the optimally-doped region (i.e. with a non-superconducting region between them, as shown here) or as a part of a ubiquitous superconducting phase (as suggested for the hole-doped $Ba_{1-x}K_xFe_2As_2$, not shown). Adapted by permission from Macmillan Publishers Ltd: Nature Physics Ref. [21], copyright (2011)

These considerations suggest that in the cuprates superconductivity is mediated by the quanta of magnetic interaction, called spin fluctuations, which necessarily exist in these compounds due to the proximity to antiferromagnetism and have a much larger energy scale than the quanta of lattice vibrations, or phonons, in conventional superconductors, thus giving rise to a much higher superconducting transition temperature. However, for electrons to form bound pairs the effective mediating interaction between them must, of course, be attractive, whereas that due to antiferromagnetic spin-fluctuations is repulsive. This dilemma can be resolved if the superconducting gap changes sign across the Fermi surface, giving rise to sign-changing coherence factors and an effective spin-fluctuation–induced attraction between electrons [22]. Sign-changing coherence factors would then imply that resonance enhancement of the inelastic neutron scattering (INS) intensity should exist in the vicinity of the binding energy of superconducting electron pairs 2Δ [23], where Δ is the superconducting energy gap. Such a resonance has indeed been observed in INS spectra and provided first experimental evidence for a spin-fluctuation pairing mechanism in high-temperature cuprate superconductors.

While the general features of the cuprates in the superconducting state have seen much consent in the community, those in the normal state, with several coexisting/competing phases, have proven much more difficult to explain and even more so to reconcile within the same theory. One of the biggest mysteries of the cuprates' normal state is the so-called 'pseudogap' or 'spin gap' phase, first discovered in nuclear magnetic resonance (NMR) measurements [13]. Since then its fingerprints have been identified with various techniques [14–17] but the complete understanding of its origin and relation to superconductivity is still lacking.

Such was the state of affairs when an entirely new family of superconducting materials was discovered in the group of Hideo Hosono in 2006 [24]. The new materials displayed structural properties quite similar to those of the cuprates: superconducting transport was found to take place in iron-pnictogen (chalcogen) planes (as compared to copper-oxygen planes in the cuprates), while intercalating/substituting atoms served as donors of free charge carriers and produced superconductivity at a certain doping level with temperatures up to 55 K [25]. Quite surprisingly, the iron-based materials turned out to possess a very similar phase diagram as well, as shown in Fig. 1.2b. Although the parent low-temperature phase was found to be an itinerant antiferromagnetic spin-density–wave (SDW) state as compared to the local antiferromagnetic state in the cuprates, the sheer proximity to antiferromagnetism provided the much higher compared to phonons energy scale thought to be a pre-requisite for high-temperature superconductivity, and, therefore, an entirely new family of superconductors for testing the predictions of already worked-out theories. There appeared only one crucial complication in these new materials that prevented straightforward knowledge extrapolation and transfer, namely, the strongly multiband character of their electronic structure as compared to the single relevant Cu-$d_{x^2-y^2}$ conduction orbital of the cuprates. To appreciate the degree of complexity introduced by this fact, it is enough to point out that computational expense (both analytical and numerical) grows exponentially with the number of relevant bands. In addition, the physical problem at hand becomes heavily overparametrized, severely limiting the amount of useful information that can be extracted from practically any study and making many inferences significantly uncertain. Under such circumstances, it becomes paramount to identify an effective low-energy theory with as low as possible dimension that nevertheless captures the essential physics of these systems. Although very challenging, this effort has already brought some fruit [26–28] but more research in this direction is certainly necessary. All of the difficulties outlined above, together with the absence of high-quality single-crystalline materials, dominated the field in the first years after the discovery of iron pnictides. Gradually, as higher-quality single crystals became available to the community, more reliable data and analysis emerged and led towards a general consensus on the properties of the superconducting and later the normal state, which is taking shape at this time. In respect of the superconducting state, for instance, the symmetry of the order parameter has been suggested to be $s_{\pm}$-type, with the superconducting gap changing sign between different sheets of the Fermi surface, and this hypothesis is consistent with a large body of experimental data for several key compounds of different classes in the iron-pnictide family, such as optimally doped with holes $BaFe_2As_2$ [29–32] and binary FeSe doped with tellurium [33]. There has been significant experimental evidence that the spin-fluctuation pairing mechanism is at the heart of superconductivity also in the iron-pnictide materials [34], even though it doesn't lead to the same symmetry of the order parameter. Many systematic studies are being carried out to extend this understanding of the superconducting state to the entire phase diagram of these materials. Also the normal state has drawn much attention and, very recently, some light has been shed upon the issue of the pseudogap [35], nematicity [36–40], and first- and second-order quantum phase transitions [37, 41–43]. This integral international research effort, together with the paradigm and experimental and theoretical techniques developed during the years of the cuprate research, as well as much improved power of modern computers, have pushed our understanding of the iron pnictides to practically same level as that of the cuprates in under 4 years as compared to more than two decades. Given our steadily improving understanding of the underpinnings of high-temperature superconductivity, we have started the transition from studying existing materials discovered mostly by chance to consciously *designing* novel high-temperature superconductors by reproducing and enhancing what are believed to be the essential details of the electronic structure of high-T_c materials [21] in atomically controlled heterostructures [44]. This research direction points towards the ultimate research goal—using well-understood building blocks to design superconducting materials with predefined mechanical and electrical properties for deployment in the evergrowing human civilization.

1.2 Scope of this Thesis

The work presented in this thesis is largely devoted to the optical properties of two major classes of the iron-based superconductors: potassium-doped $ThCr_2Si_2$-type iron arsenides ($Ba_{1-x}K_xFe_2As_2$) and recently discovered superconducting 245 iron selenides $A_2Fe_4Se_5$ (A = K, Rb, Cs). This work allowed us to extract and analyze the principal properties of the low-energy charge dynamics of these materials both in the superconducting and normal state, as well as to observe and characterize phase separation between antiferromagnetic insulating and paramagnetic superconducting phase in $Rb_2Fe_4Se_5$ (RFS).

The structure of this thesis is as follows. After a general introduction, we proceed to a brief overview of the relevant structural and electronic properties of the iron-based materials in Chap. 2. We introduce the generic crystallographic structure of these compounds and the peculiarities of individual classes, with a particular emphasis on the materials studied in this work. We proceed to outline the general phase diagram of the iron-based materials with respect to two external control parameters: alio- and/or isovalent substitution of atoms at different chemical positions and application of external hydrostatic or unidirectional pressure. This discussion is then followed by an overview of the electronic structure of these compounds in both the antiferromagnetic SDW and superconducting state. We review the band and orbital structure predicted by *ab initio* LDA calculations and compare them to experimental results obtained by means of ARPES, de Haas–van Alphen (quantum oscillations, QO), and other measurements. This chapter is concluded by a summary of the magnetic and superconducting properties of the iron-based compounds.

In Chap. 3 we introduce the main experimental techniques and methods used in this work: spectroscopic ellipsometry, scattering-type scanning near-field optical microscopy (s-SNOM), and low-energy muon-spin rotation (LE-μSR), as well as the Kramers-Kronig (KK) consistency analysis of ellipsometric data. Comprehensive spectroscopic ellipsometric measurements allow one to extract the full matrix of complex dielectric functions of any material with an arbitrary degree of anisotropy. At very low frequencies, in the far-infrared spectral range, the dielectric response of a superconducting material is determined by the properties of free charge carriers and exhibits significant suppression upon entering the superconducting state due to the development of a superconducting gap on the Fermi surface. The details of the spectral shape of the low-energy dielectric response are quite sensitive to the microscopic transport properties of free charge carriers, such as (elastic) impurity scattering rate, charge-carrier density, as well as the strength of the pairing interaction, which manifests itself in the inelastic scattering rate, and thus allows direct access to these electronic properties upon minimal data analysis under auspicious circumstances. Further insight into the microscopic superconducting properties of a material can be gained by applying a more or less sophisticated theory of superconductivity [6, 9, 45–48]. This analysis can provide evidence for the symmetry of the superconducting order parameter, type of the pairing interaction, its spectral characteristics etc. [49].

Interpretation of the optical data becomes significantly more complicated in the case of a phase-separated material because for a volume-averaged technique the optical response is a convolution of the inherent optical properties of the phases and strongly depends on the shape and volume fraction of the phase domains [50]. This additional complication necessitates a study of the material with a local imaging probe, capable of addressing all constituent phases separately. In this work we used optical imaging microscopy, realized in the s-SNOM technique. This approach combines the spatial resolution of an atomic-force microscope (AFM) with the spectral capabilities of an asymmetric Fourier-transform interferometer in one device, providing one with simultaneous topography and optical-contrast maps of the material under study. Given the reference response of a well-known material, optical contrast can me modeled to extract the inherent dielectric function with a spatial resolution limited only by the radius of the probing tip, typically as small as 10 nm.

To determine the magnetic properties of different neighboring phases, a local probe of magnetism is required. If material properties can be expected to show significant variation from the surface into the bulk then control of the probing depth is also required. All these requirements are met by LE-μSR. In this experimental technique a beam of energetic positive-charged muons μ^+ with a well-defined orientation of the spin are first slowed down to thermal velocities and then accelerated by a controlled electric field to achieve a certain kinetic energy, which then determines the stopping depth profile of muons in the sample. Once stopped, the muon spin evolves under the influence of the local magnetic field at the stopping site and finally decays into a positron, electron neutrino, and a muon antineutrino. The decay positrons can be easily detected with either scintillation or semiconductor time-resolved detectors and so the temporal evolution of the muon spin and, therefore, the local magnetic field at its stopping site in the material can be reconstructed.

After the description of the experimental techniques and methods used in this work, the main results are presented and discussed in Chap. 4. First of all, we report the complex dielectric function of $Ba_{1-x}K_xFe_2As_2$ at the optimal doping level $x = 0.4$ (BKFA) in the spectral range from 10 meV to 6.5 eV. We present its analysis in terms of the Drude-Lorentz and extended Drude models to extract the characteristic interband transitions of this system and determine the itinerant properties, respectively. It is found that the overall structure of the interband transitions agrees very well with first-principles calculations in the local-density approximation (LDA). The plasma frequency of free charge carriers is found to be 1.6 eV and the saturated high-energy optical scattering rate—on the order of 1,000 cm^{-1}, with inelastic scattering from a boson, whose spectrum is centered around 13 meV. Further, in the superconducting state we observe suppression of the far-infrared conductivity due to the opening of a superconducting gap and extract the value of the largest superconducting gap $\Delta \approx 10$ meV. The shape of the optical conductivity above 2Δ is markedly different from the conventional BCS shape obtained within the Mattis-Bardeen theory [46] and signals very small elastic scattering in this compound at least in one of the scattering channels. We model the superconducting optical response in the framework of the Eliashberg theory of superconductivity constrained by a large body of experimental data obtained with other techniques [30, 51–55]. Our analysis shows that BKFA is in the clean limit of the strong-coupling regime, with the spectrum of the mediating boson centered at 13 meV and essentially zero above 50 meV. These inferred spectral properties of the mediating boson are consistent with those of the resonance mode observed in the INS signal in the superconducting state of BKFA [29, 56], thus indicating strong coupling of itinerant electrons to spin fluctuations. A detailed examination of the optical-conductivity spectra in the visible spectral range revealed minute superconductivity-induced changes at energies as high as 2.5 eV, more than two orders of magnitude higher than the largest energy scale in this system—the superconducting energy gap $2\Delta \approx 20$ meV. Although the changes are very small and could only be identified by extensive averaging and temperature-modulation spectroscopy, the energy equivalent of the spectral weight lost at the transition into the superconducting state is equivalent, in the tight-binding nearest-neighbor approximation, to 0.60 meV/unit cell, of the same order of magnitude as the condensation energy of BKFA $\Delta F = 0.36$ meV/unit cell, determined from specific-heat measurements on the same compound [52].

We then proceed to the description of the optical properties of $A_2Fe_4Se_5$ ($A =$ K, Rb, Cs) superconducting materials. In this work we focused on the Rb-intercalated iron selenide $Rb_2Fe_4Se_5$ but a large amount of experimental evidence indicates that both electronic and magnetic properties of these materials show significant universality. In our study we find that the overall energetics of the interband transitions in quite similar to that in the 122 compounds and is also very well reproduced by *ab initio* calculations. The optical response exhibits a direct band gap of $\approx$0.45 eV and there are significantly more phonons in the far-infrared spectral range than expected for a tetragonal symmetry. All these features are consistent with the insulating vacancy-ordered state suggested for the 245 stoichiometry [57, 58]. Since no metallic response could be observed in this compound down to 10 meV, a rather unexpected for bulk superconductors[1] avenue opened up, namely, to study the THz dielectric response of this superconductor in the transmission configuration, given that sufficiently thin single-crystalline flakes can be obtained. We indeed succeeded in preparing such samples and obtained a transmission signal from them down to 1 meV. Our measurements provided first explicit evidence for an itinerant frequency response in this material. The analysis of this response revealed that it has a two-component character: a coherent (small scattering rate) and an incoherent (large scattering rate) component, commonly observed in iron-based materials [59–62], with a total plasma frequency of $\approx$130 meV, which is, however, more than an order of magnitude smaller than that of structurally similar BKFA (1.6 eV). This and other experimental evidence indicated that the material is phase-separated and the obtained optical conductivity represents an effective-medium response. In such a situation it becomes crucial to determine the volume fraction and the shape of all phases to extract their inherent dielectric properties. We chose to use a combination of s-SNOM and LE-μSR to address this issue. By means of s-SNOM we directly observed the phase separation in $Rb_2Fe_4Se_5$ in a form of nanoscale layering of insulating and metallic phase domains out-of-plane, with μm-sized domains in-plane. Using LE-μSR we showed that the majority phase is antiferromagnetic and constitutes about 80 % of the sample volume in the bulk, with the rest occupied by a spatially-separated superconducting phase. The control of the implantation depth of probing muons allowed us to study the dependence of the phase separation on the distance from the sample surface into the bulk. We discovered that the volume fraction of the antiferromagnetic phase drastically decreases near the sample surface to 50 %, with a concomitant reduction of the ordered moment to just a half of the bulk value. Effective-medium analysis of the dielectric response of $Rb_2Fe_4Se_5$ in the phase-separation picture has revealed that the inherent properties of the superconducting phase are, in fact, quite close to those of optimally doped 122 compounds, a conclusion further supported by other optical studies [63, 64] and the analysis of INS data [65].

Finally, in Chap. 5 the main results of this work are briefly summarized.

References

1. Hawking, S. W. (1974). Black hole explosions? *Nature, 248*, 30–31.
2. Ginzburg, V., & Landau, L. (1950). On the theory of superconductivity. *Zh. Eksp. Teor. Fiz., 20*, 1064.
3. Landau, L. D. (1965). Collected papers, 546 Oxord: Pergamon Press.
4. Cooper, L. N. (1956). Bound electron pairs in a degenerate Fermi gas. *Physical Review, 104*, 1189–1190.
5. Bardeen, J., Cooper, L. N., & Schrieffer, J. R. (1957). Microscopic theory of superconductivity. *Physical Review, 106*, 162–164.

[1] s-wave superconductors show zero conductivity below 2Δ, which implies unity reflectance. Therefore, no transmission can be observed in single-crystalline superconducting samples with a sufficiently large superconducting gap at THz frequencies.

6. Bardeen, J., Cooper, L. N., & Schrieffer, J. R. (1957). Theory of superconductivity. *Physical Review*, *108*, 1175–1204.
7. Gor'kov, L. (1959). Microscopic derivation of the Ginzburg-Landau equations in the theory of superconductivity. *Soviet Physics JETP*, *9*, 1364–1367.
8. Bednorz, J. G., & Müller, K. A. (1986). Possible high-T_c superconductivity in the Ba-La-Cu-O system. *Zeitschrift fur Physik B*, *64*, 189–193.
9. Eliashberg, G. (1960). Interactions between electrons and lattice vibrations in a superconductor. *Soviet Physics JETP*, *11*, 696.
10. Vaknin, D., et al. (1987). Antiferromagnetism in La_2CuO_{4-y}. *Physical Review Letters*, *58*, 2802–2805.
11. Freltoft, T., et al. (1987). Antiferromagnetism and oxygen deficiency in single-crystal $La_2CuO_{4-\delta}$. *Physical Review B*, *36*, 826–828.
12. Pickett, W. E. (1989). Electronic structure of the high-temperature oxide superconductors. *Reviews of Modern Physics*, *61*, 433–512.
13. Alloul, H., Ohno, T., & Mendels, P. (1989). ^{89}Y NMR evidence for a Fermi-liquid behavior in $YBa_2Cu_3O_{6+x}$. *Physical Review Letters*, *63*, 1700–1703.
14. Rossat-Mignod, J., et al. (1991). Investigation of the spin dynamics in $YBa_2Cu_3O_{6+x}$ by inelastic neutron scattering. *Physics B*, *169*, 58–65.
15. Rotter, L. D., et al. (1991). Dependence of the infrared properties of single-domain $YBa_2Cu_3O_{7-y}$ on oxygen content. *Physical Review Letters*, *67*, 2741–2744.
16. Tallon, J. L., Cooper, J. R., de Silva, P. S. I. P. N., Williams, G. V. M., & Loram, J. W. (1995). Thermoelectric power: A simple, instructive probe of high-T_c superconductors. *Physical Review Letters*, *75*, 4114–4117.
17. Loeser, A. G., et al. (1996). Excitation gap in the normal state of underdoped $Bi_2Sr_2CaCu_2O_{8+\delta}$. *Science*, *273*, 325–329.
18. Ghiringhelli, G., et al. (2012). Long-range incommensurate charge fluctuations in $(Y, Nd)Ba_2Cu_3O_{6+x}$. *Science*, *337*, 821–825.
19. Achkar, A. J., et al. (2012). Distinct charge orders in the planes and chains of ortho-III-ordered $YBa_2Cu_3O_{6+\delta}$ superconductors identified by resonant elastic X-ray scattering. *Physical Review Letters*, *109*, 167001.
20. Kohsaka, Y., et al. (2012). Visualization of the emergence of the pseudogap state and the evolution to superconductivity in a lightly hole-doped Mott insulator. *Nature Physics*, *8*, 534–538.
21. Basov, D. N., & Chubukov, A. V. (2011). Manifesto for a higher T_c. *Nature Physics*, *7*, 272–276.
22. Scalapino, D. J., Loh, E., & Hirsch, J. E. (1986). d-wave pairing near a spin-density-wave instability. *Physical Review B*, *34*, 8190–8192.
23. Tinkham, M. (1995). *Introduction to superconductivity* (2nd ed.). New York: McGraw-Hill.
24. Kamihara, Y., et al. (2006). Iron-based layered superconductor: LaOFeP. *Journal of the American Chemical Society*, *128*, 10012–10013.
25. Zhi-An, R., et al. (2008). Superconductivity at 55 K in iron-based F-doped layered quaternary compound $Sm[O_{1-x}F_x]FeAs$. *Chinese Physics Letters*, *25*, 2215.
26. Hu, J., & Hao, N. (2012). S_4 symmetric microscopic model for iron-based superconductors. *Physical Review X*, *2*, 021009.
27. Ma, T., Lin, H.-Q., & Hu, J. (2013). Quantum Monte-Carlo study of a dominant s-wave pairing symmetry in iron-based superconductors. *Physical Review Letters*, *110*, 107002.
28. Hao, N., Wang, Y. & Hu, J. (2012). Oriented gap opening in the magnetically ordered state of iron-pnicitides: an impact of intrinsic unit cell doubling on the Fe square lattice by As atoms. arXiv:1207.6798 (unpublished).
29. Christianson, A. D., et al. (2008). Unconventional superconductivity in $Ba_{0.6}K_{0.4}Fe_2As_2$ from inelastic neutron scattering. *Nature*, *456*, 930.
30. Evtushinsky, D. V., et al. (2009). Momentum dependence of the superconducting gap in $Ba_{1-x}K_xFe_2As_2$. *Physical Review B*, *79*, 054517.
31. Xu, Y.-M. et al. (2011). Observation of a ubiquitous three-dimensional superconducting gap function in optimally doped $Ba_{0.6}K_{0.4}Fe_2As_2$. *Nature Physics*, *7*, 198.

32. Evtushinsky, D. V. et al. (2012). Strong pairing at iron $3d_{xz,yz}$ orbitals in hole-doped $BaFe_2As_2$. arXiv:1204.2432 (unpublished).
33. Hanaguri, T., Niitaka, S., Kuroki, K., & Takagi, H. (2010). Unconventional s-wave superconductivity in Fe(Se,Te). *Science, 328*, 474.
34. Paglione, J., & Greene, R. L. (2010). High-temperature superconductivity in iron-based materials. *Nature Physics, 6*, 645.
35. Moon, S. J., et al. (2012). Infrared measurement of the pseudogap of P-doped and Co-doped high-temperature $BaFe_2As_2$ superconductors. *Physical Review Letters, 109*, 027006.
36. Chu, J.-H., et al. (2010). In-plane resistivity anisotropy in an underdoped iron arsenide superconductor. *Science, 329*, 824–826.
37. Chu, J.-H., Kuo, H.-H., Analytis, J. G., & Fisher, I. R. (2012). Divergent nematic susceptibility in an iron arsenide superconductor. *Science, 337*, 710–712.
38. Kasahara, S., et al. (2012). Electronic nematicity above the structural and superconducting transition in $BaFe_2(As_{1-x}P_x)_2$. *Nature, 486*, 382–385.
39. Kuo, H.-H., et al. (2012). Magnetoelastically coupled structural, magnetic, and superconducting order parameters in $BaFe_2(As_{1-x}P_x)_2$. *Physical Review B, 86*, 134507.
40. Wang, A. F., et al. (2013). A crossover in the phase diagram of $NaFe_{1-x}Co_xAs$ determined by electronic transport measurements. *New Journal of Physics, 15*, 043048.
41. Hashimoto, K., et al. (2012). A sharp peak of the zero-temperature penetration depth at optimal composition in $BaFe_2(As_{1-x}P_x)_2$. *Science, 336*, 1554–1557.
42. Arsenijević, S. et al. (2012). Spin fluctuation driven enhancement of thermopower near the quantum critical point in $Ba(Fe_{1-x}Co_x)_2As_2$. arXiv:1206.2938 (unpublished).
43. Kitagawa, S., Ishida, K., Nakamura, T., Matoba, M., & Kamihara, Y. (2012). Ferromagnetic quantum critical point in heavy-fermion iron oxypnictide $Ce(Ru_{1-x}Fe_x)PO$. *Physical Review Letters, 109*, 227004.
44. Logvenov, G., Gozar, A., & Bozovic, I. (2009). High-temperature superconductivity in a single copper-oxygen plane. *Science, 326*, 699–702.
45. Mattis, D. C., & Bardeen, J. (1958). Theory of the anomalous skin effect in normal and superconducting metals. *Physical Review, 111*, 412–417.
46. Zimmermann, W., Brandt, E., Bauer, M., Seider, E., & Genzel, L. (1991). Optical conductivity of BCS superconductors with arbitrary purity. *Physical C, 183*, 99.
47. Nam, S. B. (1967). Theory of electromagnetic properties of superconducting and normal systems. *I Physical Review, 156*, 470.
48. Nam, S. B. (1967). Theory of electromagnetic properties of strong-coupling and impure superconductors. *II. Physical Review, 156*, 487.
49. Carbotte, J. P. (1990). Properties of boson-exchange superconductors. *Review Modern Physics, 62*, 1027–1157.
50. Choy, T. C. (1999). *Effective medium theory : principles and applications*. Oxford: Oxford University Press.
51. Evtushinsky, D. V., et al. (2009). Momentum-resolved superconducting gap in the bulk of $Ba_{1-x}K_xFe_2As_2$ from combined ARPES and μSR measurements. *New Journal of Physics, 11*, 055069.
52. Popovich, P., et al. (2010). Specific heat measurements of $Ba_{0.68}K_{0.32}Fe_2As_2$ single crystals: evidence for a multiband strong-coupling superconducting state. *Physical Review Letters, 105*, 027003.
53. Shan, L., et al. (2011). Evidence of multiple nodeless energy gaps in superconducting $Ba_{0.6}K_{0.4}Fe_2As_2$ single crystals from scanning tunneling spectroscopy. *Physical Review B, 83*, 060510.
54. Shan, L. et al. (2011). Observation of ordered vortices with Andreev bound states in $Ba_{0.6}K_{0.4}Fe_2As_2$. *Nature Physics, 7*, 325.
55. Shan, L., et al. (2012). Evidence of a spin resonance mode in the iron-based superconductor $Ba_{0.6}K_{0.4}Fe_2As_2$ from scanning tunneling spectroscopy. *Physical Review Letters, 108*, 227002.
56. Zhang, C. et al. (2011). Neutron scattering studies of spin excitations in hole-doped $Ba_{0.67}K_{0.33}Fe_2As_2$ superconductor. *Science Report, 1*, 115.

57. Bacsa, J., et al. (2011). Cation vacancy order in the $K_{0.8+x}Fe_{1.6-y}Se_2$ system: Five-fold cell expansion accommodates 20% tetrahedral vacancies. *Chemical Sciences*, *2*, 1054.
58. Yan, X.-W., Gao, M., Lu, Z.-Y., & Xiang, T. (2011). Ternary iron selenide $K_{0.8}Fe_{1.6}Se_2$ is an antiferromagnetic semiconductor. *Physical Review B*, *83*, 233205.
59. Wu, D., et al. (2009). Effects of magnetic ordering on dynamical conductivity: Optical investigations of $EuFe_2As_2$ single crystals. *Physical Review B*, *79*, 155103.
60. Nakajima, M., et al. (2010). Evolution of the optical spectrum with doping in $Ba(Fe_{1-x}Co_x)_2As_2$. *Physical Review B*, *81*, 104528.
61. Barišić, N., et al. (2010). Electrodynamics of electron-doped iron pnictide superconductors: Normal-state properties. *Physical Review B*, *82*, 054518.
62. Cheng, B., et al. (2012). Electronic properties of 3d transitional metal pnictides: A comparative study by optical spectroscopy. *Physical Review B*, *86*, 134503.
63. Wang, C. N., et al. (2012). Macroscopic phase segregation in superconducting $K_{0.73}Fe_{1.67}Se_2$ as seen by muon spin rotation and infrared spectroscopy. *Physical Review B*, *85*, 214503.
64. Homes, C. C., Xu, Z. J., Wen, J. S., & Gu, G. D. (2012). Effective medium approximation and the complex optical properties of the inhomogeneous superconductor $K_{0.8}Fe_{2-y}Se_2$. *Physical Review B*, *86*, 144530.
65. Friemel, G., et al. (2012). Reciprocal-space structure and dispersion of the magnetic resonant mode in the superconducting phase of $Rb_xFe_{2-y}Se_2$ single crystals. *Physical Review B*, *85*, 140511.

Chapter 2
Iron-Based Superconductors

It is a capital mistake to theorize before one has data. Insensibly one begins to twist facts to suit theories, instead of theories to suit facts

—Sherlock Holmes, *A Scandal in Bohemia*

Two decades into the intensive study of the cuprate superconductors, the condensed-matter community got stirred up once again when another completely different family of superconductors was discovered by the group of Hideo Hosono in 2006 [1]. The Japanese group reported observation of a superconducting transition in LaFePO at a relatively low temperature of ~4 K. This original discovery received certain but limited attention from the community. The general excitement came 2 years later when the same group reported superconductivity at a temperature of 26 K, higher that that of most conventional superconductors, in a closely related compound $LaFeAsO_{1-x}F_x$ at a doping level of $x = 0.12$ [2], with the parent compound LaFeAsO being non-superconducting at routinely attainable cryogenic temperatures. This latter discovery gave rise to the explosive growth of research of these materials all over the world, which lead to reports of high-temperature superconductivity[1] in several new classes of compounds in this family, such as $SmFeAsO_{0.9}F_{0.1}$ [4] ($T_c \approx 55$ K) and $Ba_{0.6}K_{0.4}Fe_2As_2$ [5] ($T_c \approx 38$ K).

The iron-based materials are classified based on their crystallographic structure. Separate classes are usually denoted by the chemical formula of the parent, often non-superconducting, compounds, e.g. 1111 for the parent compounds REFeAsO, RE = La, Sm, Gd, ... or 122 for $BaFe_2As_2$. The most relevant and studied classes of iron-based compounds are shown in Fig. 2.1a. A common unifying feature to all these

[1] In this work we adopt the generalized definition of 'high-temperature superconductors' as materials with the superconducting transition temperature above 28 K (T_c of most conventional superconductors falls below this value. A well-known exception is MgB_2 with $T_c = 39$ K [3], which is a conventional, albeit a two-band superconductor). This definition differs from the historical, industrially important, notion of high-temperature superconductors as materials with T_c above the boiling point of liquid nitrogen (77 K).

A. Charnukha, *Charge Dynamics in 122 Iron-Based Superconductors*, Springer Theses, DOI: 10.1007/978-3-319-01192-9_2,

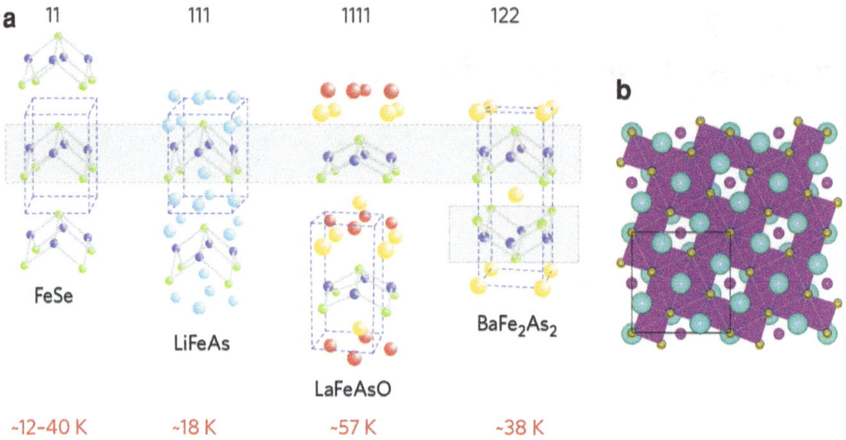

Fig. 2.1 **a** Several representative classes of the iron-based materials. The *numbers* at the *top* denote the chemical formula of the compound underneath (for instance, LiFeAs is a 111 compound), while the *temperatures* at the *bottom* specify either the superconducting transition temperature of the material itself or the highest T_c obtained by doping or substitution in a material of that type. The common structural unit of all iron-based compounds is the Fe–As tetrahedral layer (*gray areas*), with the adjacent layers either co-aligned out-of-plane or alternating in orientation. Adapted by permission from Macmillan Publishers Ltd: Nature Physics [6], copyright (2009). **b** In-plane crystallographic structure of 245 iron selenides $A_2Fe_4Se_5$ (A = K, Rb, Cs). Iron vacancies are depicted as *purple spheres*, selenium sites as *golden* and alkaline-metal sites as *large blue spheres*. The $\sqrt{5} \times \sqrt{5}$ enlargement of the unit cell is evident. The out-of-plane crystallographic structure is essentially identical to that of the 122 compounds. Reproduced by permission of The Royal Society of Chemistry from Ref. [7]

classes is the basic structural building block present in all iron-based materials—the Fe–As layer. This layer is composed of a square lattice of iron atoms (blue spheres in Fig. 2.1a) and an interpenetrating lattice of As atoms (green spheres) alternately displaced to above and below the plane of the iron lattice, forming almost regular tetrahedra. The difference between the classes can be reduced to the structure of the blocking layer sandwiched in between the adjacent Fe–As layers (none for the 11, monatomic for the 111 and 122, and diatomic for the 1111 compounds) and the orientation of the adjacent Fe–As layers with respect to each other (all oriented in the same direction for the 11, 111, and 1111 materials and with the alternating orientation for the 122 compounds). This work is focused on the materials in the 122 class or its generalization that encompasses the so-called vacancy-ordered 245 iron selenides $A_2Fe_4Se_5$ (A = K, Rb, Cs), whose general crystallographic structure is identical to that of 122 systems ($A_2Fe_4Se_5 = A_{0.8}Fe_{1.6}Se_2 \in A_{1-x}Fe_{2-y}Se_2$) but with ordered vacancies in both the intercalating-atom (x) and iron (y) sublattices (Fig. 2.1b).

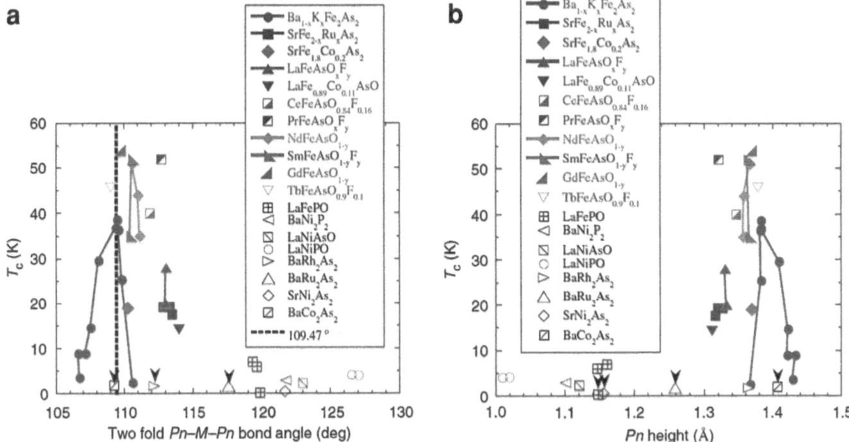

Fig. 2.2 Dependence of the superconducting transition temperature of a set of representative iron-based superconductors on the Pn–Fe–Pn bond angle (**a**), where Pn is a pnictogen, and on the pnictogen height (**b**). Adapted by permission from Taylor and Francis Ltd. (http://www.informaworld.com): Ref. [8], copyright (2010)

2.1 Crystallographic and Reciprocal-Space Structure

As it was shown in Fig. 1.2b, the iron-based materials exhibit three principal phases: normal (white areas), antiferromagnetic SDW (red), and superconducting (yellow). The nematic phase, shown as blue areas, is a more exotic and elusive phase, which shows no magnetism but exhibits lower point-group symmetry than the normal metallic phase. We will elaborate on the nature of this phase later on in Sect. 2.2. Since in the study reported in this thesis we have concentrated on the 122 class of the iron-based superconductors, predominantly the properties of materials pertaining to this class will be reviewed here. A more general account, as well as an extensive comparison of various structural, electronic, magnetic, and superconducting properties of several representative materials of this family can be found in excellent reviews on the topic, for instance, in Ref. [8]. A good early topical review of all current iron-based materials was given in issues 9–12 of *Physica C: Superconductivity*, *469*, 313–674 (2009).

The prototypical compound of the 122 class of iron-based superconductors, and also the one that produces the highest superconducting transition temperature of 38 K upon potassium substitution, is $Ba_{1-x}K_xFe_2As_2$ (an extensive comparison of the maximum superconducting transition temperatures and other superconducting properties of many iron-based superconductors in the 122 and other classes can be found in Ref. [9]). Other compounds in this class have been synthesized by replacing Ba with other alkali-earth metals (Sr and Ca) intercalating between the iron-arsenic layers and have properties and the phase diagram very similar to those of $Ba_{1-x}K_xFe_2As_2$. The parent (undoped) compound of the latter is $BaFe_2As_2$ and crystallizes in a $ThCr_2Si_2$-type structure (see Fig. 2.1) with a tetragonal crystallographic symmetry in the $I4/mmm$ space group and two formula units per simple tetragonal unit cell [10]. The lattice parameters in the tetragonal phase at room temperature have been found to be $a = 3.9625(1)$ Å and $c = 13.0168(3)$ Å [10]. Upon cooling this compound undergoes almost coincident structural and magnetic phase transitions at $T_s \approx T_N \approx 136$ K into an orthorhombic (rotated around the principal tetragonal axis by 45° with respect to the tetragonal unit cell) antiferromagnetic SDW state of space group $Fmmm$ with the lattice constants $a = 5.6146(1)$ Å, $b = 5.5742(1)$ Å, and $c = 12.9453(1)$ Å [10].

This compound, like all iron-based materials due to their common iron-arsenic (or, more generally, iron-pnictoden/chalcogen) structural unit, exhibits tetrahedral coordination of Fe and As. Iron ions form a square lattice, with As ions occupying the central planar positions alternately above or below the Fe plane. The main parameter characterizing such coordination is the pnictogen height or, equivalently, the As–Fe–As bond angle. The superconducting properties of the iron-based materials have been shown to display a certain degree of correlation with these two parameters, in particular, the superconducting transition temperature tends to maximize close to the perfect tetrahedral geometry, i.e. for an As–Fe–As bond of $\arccos(-1/3) \approx 109.47°$, as shown in Fig. 2.2a and, analogously for the pnictogen height, in Fig. 2.2b. Recently, first evidence for a *local* correlation between the pnictogen/chalcogen height and the rigidity of the superconducting condensate has been provided by scanning tunneling microscopy/spectroscopy (STM/STS) on FeSe thin films grown by molecular beam epitaxy [11]. Although the superconducting properties of the compounds in Fig. 2.2 obtained by aliovalent substitution are affected not only by structural distortions but also by direct chemical doping, there is significant experimental evidence that it is the structurally driven modification of the Fermi surface, rather than charge-carrier doping, which optimizes the superconducting transition temperature [12, 13].

Since some experimental probes are sensitive predominantly to Fe, rather than to all constituents, three different unit cells can often be encountered in the literature and are shown in Fig. 2.3a–c, most importantly: the 4-Fe simple tetragonal unit cell with two formula units per unit cell (Fig. 2.3a), the 1-Fe unit cell based on the Fe sublattice, which would be realized physically if all pnictogen atoms were located in the Fe plane (Fig. 2.3b), and the 2-Fe unit cell, which is the real primitive unit cell of the iron-based materials (Fig. 2.3c). These unit cells generate corresponding reciprocal unit cells, or the first Brillouin zones [14], as also shown in Fig. 2.3.

2.2 Phase Diagram

The phase diagram of all iron-based materials shows many common features and can therefore be cast into a general schematic form shown in Fig. 1.2b. The parent compounds in this family are semimetals at room temperature and exhibit an

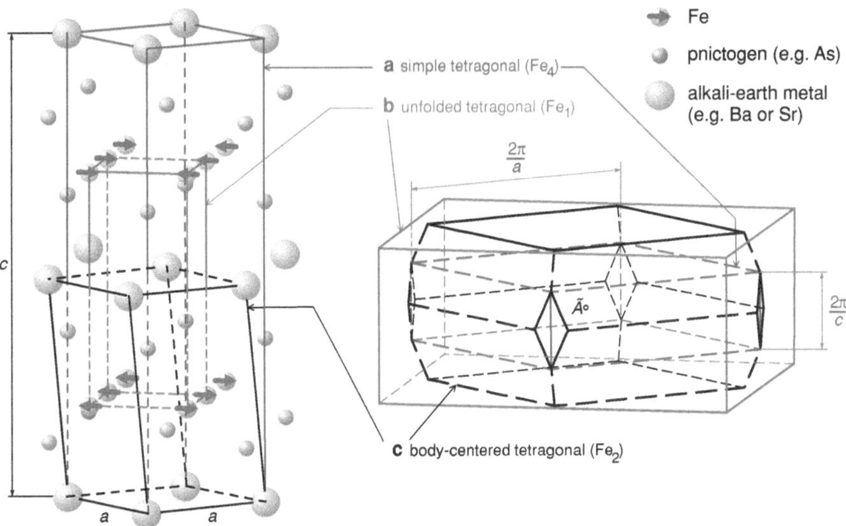

Fig. 2.3 Three different unit cells of the 122 iron pnictides: **a** the 4-Fe simple tetragonal unit cell with two formula units per unit cell, **b** the 1-Fe unit cell based on the Fe sublattice alone, and **c** the 2-Fe unit cell, which is the real primitive unit cell of the iron-based materials. Shown on the *right* are the first Brillouin zones corresponding to these unit cells. Reprinted figure with permission from Ref. [15]. Copyright (2010) by the American Physical Society. Original material obtained from and modified with permission of the author

antiferromagnetic SDW phase transition at lower temperatures.[2] In the rare-earth-based compounds of the 1111 class, a complex interplay of itinerant Fe and localized rare-earth moments gives rise to a sequence of (anti)ferromagnetic phase transitions [16–18], which, however, do not seem to bear significantly on the nature of superconducting pairing. What appears to be of much importance for superconductivity in these materials, and this observation correlates strongly with that in the cuprates, is the proximity to the antiferromagnetism of itinerant electrons. This proximity gives rise to strong antiferromagnetic spin fluctuations once the long-range order has been suppressed, which are currently believed to be the most plausible candidate for the superconducting pairing mediator [9, 19–25].

There are three main means of suppressing long-range antiferromagnetic order and inducing superconductivity in the iron-based compounds: aliovalent substitution [doping, e.g. $Ba_{1-x}K_xFe_2As_2$, $Ba(Fe_{1-x}Co_x)_2As_2$; see Fig. 2.4], isovalent substitution [modification of the crystallographic structure, e.g. $BaFe_2(As_{1-x}P_x)_2$, $Ba(Fe_{1-x}Ru_x)_2As_2$], and application of external pressure. All these methods have

[2] with the exception of the iron-selenide compounds with vacancy ordering, which are dominated by an antiferromagnetic semiconducting phase at room temperature. Their complete classification is complicated by the fact that no single-phase superconducting compounds have been synthesized up to date and even the identification of the parent compound of this class of materials appears problematic.

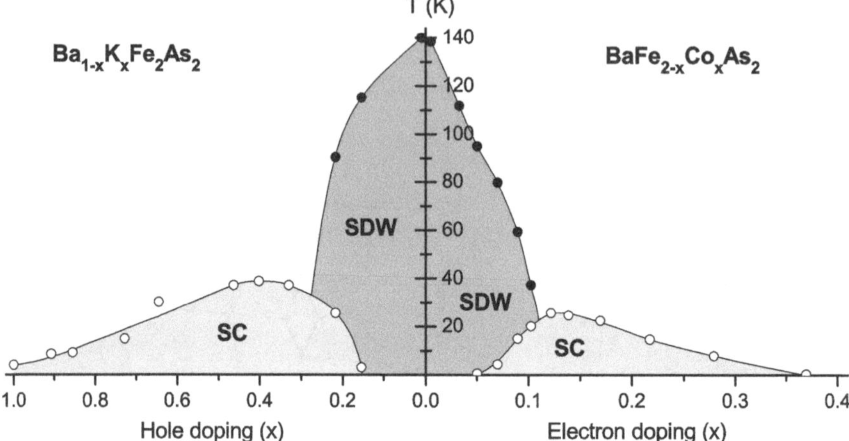

Fig. 2.4 Phase diagram of $Ba_{1-x}K_xFe_2As_2$ (*left*) and $Ba(Fe_{1-x}Co_x)_2As_2$ (*right*). Data from Refs. [34, 35], respectively. In both series of compounds the overlap region of the SDW and superconducting phase has been found to exhibit the microscopic coexistence of the two order parameters in the single crystals of sufficiently high quality [28, 31]

been found to produce very similar results, namely, a quick suppression of long-range antiferromagnetism with subsequent induction of superconductivity in a characteristic dome-shaped region of the phase diagram [8, 26]. In certain materials a further increase of external pressure leads to a re-entrant superconducting phase with an even higher superconducting transition temperature [27].

In those iron-based compounds showing overlapping antiferromagnetic and superconducting regions of the phase diagram (see Figs. 2.2 and 2.4) the issue of phase separation/coexistence has been a topic of intense debate. At the dawn of the iron-pnictide research it was believed, for instance, that some materials, such as $Ba(Fe_{1-x}Co_x)_2As_2$, show microscopic coexistence of antiferromagnetism and superconductivity [28], whereas others, such as $Ba_{1-x}K_xFe_2As_2$, undergo intrinsic phase separation [29]. Recently, experimental evidence in favor of microscopic phase coexistence in $Ba_{1-x}K_xFe_2As_2$ as well as in $Ba(Fe_{1-x}Ru_x)_2As_2$ single crystals of sufficiently high quality has emerged [30–32], proving that phase separation is an extrinsic effect, at least in the 122 iron pnictides. Given that both phases coexist microscopically and homogeneously throughout the bulk of the crystal their properties must be strongly intertwined, making it possible to study the properties of one via the other. One important example in the iron-pnictide research is the type of the quantum (zero-temperature) phase transition at the boundary of the antiferromagnetic phase in the superconducting dome. If this point of the phase diagram were a quantum critical point, i.e. if the quantum phase transition from the coexistence state to the pure superconducting state were a second-order quantum phase transition, certain physical quantities would diverge. A recent study of one of the iron-pnictide compounds with phase coexistence, $BaFe_2(As_{1-x}P_x)_2$, has indeed revealed a strong

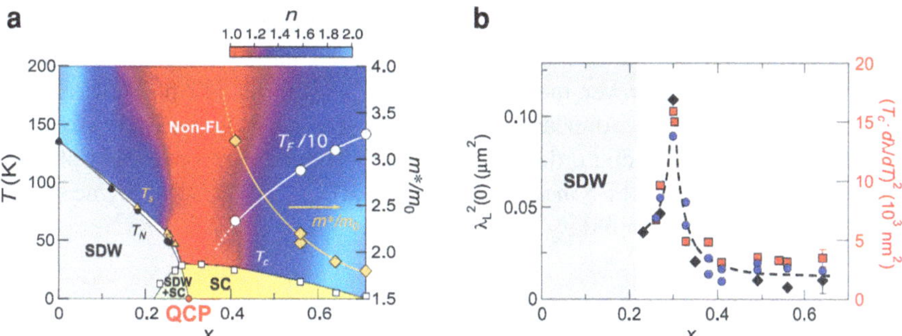

Fig. 2.5 **a** Phase diagram of $BaFe_2(As_{1-x}P_x)_2$. A quantum critical point separates the zero-temperature phase of coexisting antiferromagnetic SDW and superconductivity from the pure superconducting phase. The non-Fermi-liquid regime (*red region*) results from finite-temperature critical fluctuations due to the underlying quantum critical point. **b** Doping dependence of the zero-temperature London penetration depth (*left vertical axis*) and its temperature derivative (*right vertical axis*). From Ref. [33]. Reprinted with permission from AAAS

peak in the doping dependence of the zero-temperature London penetration depth $\lambda_L(0)$ at the doping level corresponding to the boundary of the coexistence phase, as shown in Fig. 2.5, thus confirming the second-order character of the corresponding phase transition [33]. The fact that the maximum of the superconducting transition temperature and the divergence of $\lambda_L(0)$ occur at the same doping level underlines the importance of critical antiferromagnetic spin fluctuations for superconductivity in the iron-based materials and reveals their dual role: critical enhancement of these spin fluctuations leads to the stronger binding of electrons into Cooper pairs and thus a higher T_c, while the concomitant increase in the scattering of electrons from these excitations makes the Cooper pairs heavier and, therefore, their screening of external magnetic fields less effective, hence the increase of the penetration depth.

Needless to say, the (thermal) phase transition from the normal to the superconducting state is of second order in the absence of an external magnetic field [36]. The transition from the normal to the antiferromagnetic state, coupled to the structural transition from the high-temperature tetragonal to the low-temperature orthorhombic phase in the 122 materials, is strongly first-order [37–41], with the exception of $BaFe_2As_2$, which is believed to exhibit a weakly first-order phase transition [41]. 'Weakly' in this context implies that rather than being a conventional sharp discontinuous transition it extends over a sizable temperature region, giving rise to a finite resistivity anisotropy above the antiferromagnetic transition proper [41, 42].

Finally, the early discovery of resistivity anisotropy above the nominal structural transition in the parent and Co-doped $BaFe_2As_2$ [42, 43] has raised the question of the possible existence of another phase, at least in the 122 iron pnictides: a nematic phase, which breaks the $C4$ rotational symmetry of the high-temperature tetragonal phase in a second-order phase transition. Since then, the issue of nematicity in the iron-based materials has received much attention. This new phase has been both confirmed in the Co-doped compound [44] and discovered using magnetic-torque measurements in

$BaFe_2(As_{1-x}P_x)_2$ [45]. The occurrence of the nematic phase in these compounds can be explained by anisotropic spin fluctuations above the antiferromagnetic transition temperature [46, 47]. However, the fact that in Ref. [45] the nematic phase was found to persist well into the overdoped regime of $BaFe_2(As_{1-x}P_x)_2$, far away from the long-range antiferromagnetic order, indicates that orbital anisotropy might play an important role. Recent ARPES measurements on $Ba(Fe_{1-x}Co_x)_2As_2$ provide strong support to this hypothesis [48].

2.3 Electronic Band Structure

The electronic band structure of the iron-based materials near the Fermi level is largely defined by the outer orbitals of the iron (3d) and pnictogen/chalcogen (4p) atoms. The details of the crystallographic structure and charge-carrier doping determine the exact shape of the bands and the relative location of the chemical potential and thus the geometry and topology of the Fermi surface. All iron-based compounds have been found to be multiband systems, with several separate sheets of the Fermi surface within the Brillouin zone [8, 26]. More often than not, both hole- and electron-type pockets are present, with up to a total of five. The most direct way to determine the Fermi surface of a compound is by means of ARPES. By detecting emitted electrons with the energy equal to the Fermi energy, synchrotron-based ARPES is capable of mapping the Fermi surface in the entire Brillouin zone [49]. When carried out in the superconducting state, ARPES measurements provide detailed information about the momentum-dependence of the superconducting gap on all Fermi surfaces where it can be resolved (the ultimate resolution limit is instrument-dependent, with the state-of-the-art experiments discerning superconducting gaps as small as 3 meV). When sufficiently clean single-crystalline materials are available, the Fermi surface can also be probed by various QO measurements [50].

Both ARPES and QO techniques allow for a comparison of the experimentally inferred band structure with the predictions of first-principles calculations. The information extracted from this comparison is twofold. Firstly, the overall adequacy of the ab initio description of a given class of materials can be assessed by comparing the gross band dispersions in the experimentally accessible energy range; the comparison of the widths of the bands tells one how strong high-energy correlations are in the system: the more compressed the experimental bands appear with respect to their calculated counterparts, the stronger the high-energy correlations are [51]. Secondly, the investigation of the deviations of the low-energy dispersion (i.e. in the vicinity of the Fermi level) from the prediction quantifies the strength of the coupling of free charge carriers to various low-energy excitations in the system, such as phonons, spin, charge and orbital fluctuations, etc. [49]. Given that the electron–electron interaction mediated by at least one type of these excitations is effectively attractive (intraband phonon-mediated interaction is always attractive), it will lead to the creation of Cooper pairs and thus the emergence of superconductivity. Since the effective mass of itinerant charge carriers in a band is defined as the second momentum derivative of that band's dispersion at the Fermi level, any deviation of the band dispersion from the theoretical prediction implies a different effective mass, the effect called 'effective-mass renormalization'. In principle, both ARPES and QO measurements can determine this mass renormalization but in practice the latter have a much higher accuracy in this regard. In ARPES measurements, on the other hand, one can utilize different polarization states of the incident light to extract the orbital characters of the bands, which can also be compared with theoretical predictions, providing another benchmark for the adequacy of the approximations used in the calculations. More importantly, the knowledge of the orbital structure of the bands provides essential input for the more involved theories of various physical phenomena, e.g. the microscopic theory of superconductivity.

After this general discussion let us consider several concrete examples elucidating the electronic band structure of the 122 iron-based materials. The major common electronic properties of the former are well represented by the $Ba_{0.6}K_{0.4}Fe_2As_2$ compound. Figure 2.6a shows the electronic band structure of this material obtained with LDA [52] using the experimental values of the lattice constants and atomic positions in the unit cell determined by means of a Rietveld fit of powder X-ray diffraction data [10]. The different colors of the bands are used to illustrate the different orbital contributions dominating each band. It is immediately evident that

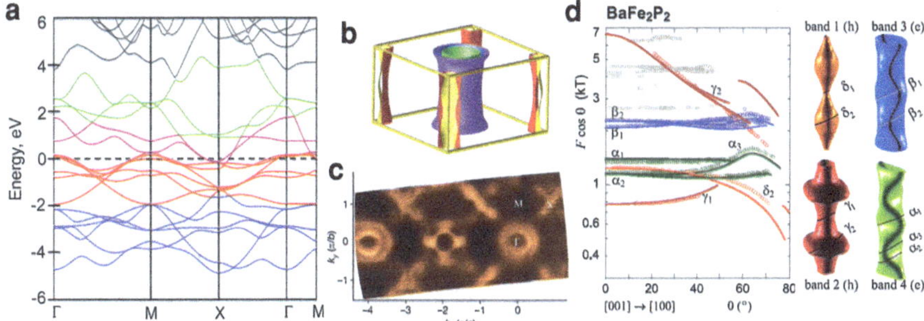

Fig. 2.6 **a** Electronic band structure of $Ba_{0.6}K_{0.4}Fe_2As_2$ obtained in LDA calculations using the experimentally determined lattice constants and atomic positions. The colors represent different dominant orbital contributions: Fe-d (*red*), As-p (*blue*), Ba-d (*black*); *green lines* denote strongly hybridized Fe-d and As-p orbitals and *purple lines* depict the Fe-d orbitals that give rise to the electron pockets of the Fermi surface. Adapted by permission from Macmillan Publishers Ltd: Nature Communications Ref. [52], copyright (2011). **b** Fermi surface of $Ba_{0.6}K_{0.4}Fe_2As_2$ predicted by ab initio calculations. Reprinted from Ref. [53], copyright (2009), with permission from Elsevier. **c** Its experimental counterpart obtained with ARPES. Reprinted by permission from Macmillan Publishers Ltd: Nature Ref. [54], copyright (2009). **d** Angular dependence of the QO frequencies extracted from the data obtained on a $BaFe_2P_2$ compound (open circles) superimposed onto the prediction of a first-principles calculation for the same compound (*solid lines*; the calculated bands have been shifted to match the experimental data as discussed in the text). (*Far right*) Partial Fermi surfaces as predicted by theory with the assignment of the QO frequencies from the *left panel*. Reprinted figure with permission from Ref. [55]. Copyright (2011) by the American Physical Society

the electronic structure in the vicinity of the Fermi energy (± 2 eV) is dominated by either by Fe-d (red, purple) or strongly hybridized Fe-d and As-p orbitals (green), as maintained in the introduction of this section. Deeper orbitals have mostly As-p character (blue lines), while the contribution of Ba-d orbitals becomes significant only above $\sim$4 eV. The predominantly Fe-d-character bands crossing the Fermi level are holelike at the Γ point and electronlike at the X point of the 1-Fe Brillouin zone and the crossings determine the complex multiband Fermi surface of this compound shown in Fig. 2.6b. It is remarkable that, similarly to the copper–oxygen planes in the cuprates, the iron–pnictogen/chalcogen planes in the iron-based materials are only weakly coupled, making the Fermi surface two-dimensional to a large extent, with only a minor warping of its quasicylindrical sheets. Such two-dimensionality of the electronic structure is now widely believed to be one of the essential prerequisites for high-temperature superconductivity [56], mostly due to the drastic enhancement of electronic instabilities and fluctuations in lower dimensions.

The overall geometry and topology of the Fermi surface predicted by first-principles calculations as described above have been found largely consistent with experiment. All techniques sensitive to the shape of the Fermi surface have confirmed its multiband and quasi-two-dimensional character. For example, Fig. 2.6c shows the in-plane Fermi surface of $Ba_{0.7}K_{0.3}Fe_2As_2$ mapped out with ARPES. Two concentric circular pockets are clearly visible at the Γ point, along with a complex 'propeller' structure at the X point of the 1-Fe Brillouin zone. The latter consists of a central electronlike pocket surrounded by four 'blades', which have a holelike character and supposedly result from high-temperature Fermi-surface reconstruction due to a nesting instability (good match in their geometrical shape and size) between one of the circular hole pockets at the Γ point and an electron pocket at the X point of the 1-Fe magnetic Brillouin zone (M point of the 2-Fe Brillouin zone corresponding to the real chemical unit cell) [54]. Thus the topology of the 'propeller' structure does indeed require two electronlike pockets at the X point of the magnetic Brillouin zone, as predicted by theory (see Fig. 2.6b). Subsequent low-temperature ARPES measurements in the superconducting state of $Ba_{0.6}K_{0.4}Fe_2As_2$ have revealed three superconducting gaps centered around the Γ point, which requires three circular Fermi pockets in full agreement with ab initio calculations. In general, the highest-resolution reconstruction of the Fermi surface is done based on ARPES data obtained in the superconducting state. Since superconductivity gaps the entire Fermi surface symmetrically above and below (with possible nodes, i.e. points in the Brillouin zone where the superconducting gap vanishes), it is a trivial task to extract the shape of the latter from the complete momentum-dependence of the superconducting gap(s) in the Brillouin zone, which is routinely extracted from ARPES data [57, 58]. In addition, low-temperature ARPES measurements have a much higher resolution compared to those at room temperature due to the strong suppression of thermal broadening. Therefore, we postpone a more detailed review of the Fermi surface of $Ba_{0.6}K_{0.4}Fe_2As_2$ extracted from ARPES measurements until Sect. 2.5.

As outlined above, measurements of QO is a very powerful tool of condensed-matter physics and provides access to the essential properties of quasiparticle dynamics. The practical feasibility of these measurements, however, requires very clean materials [14], which means that high doping levels are detrimental to the successful application of this technique. Unfortunately, almost all optimally doped iron-based superconducting materials showing the highest superconducting transition temperature are relatively heavily doped/substituted. In addition, QO can only be observed in a metallic compound with a finite Fermi surface. Therefore, superconductivity must be fully suppressed by applying a magnetic field above the upper critical field to recover the normal state and allow for QO measurements. This fact makes such experiments very challenging to perform on optimally doped iron-based superconductors because many of them show upper critical fields above 150 T [8]. Such magnetic fields are inaccessible even to the most powerful state-of-the-art pulsed magnets. The above limitations clarify why the overwhelming majority of QO measurements have been carried out on the end members of various doping series. Since all undoped iron-based compounds are itinerant antiferromagnets, their low-temperature Fermi surface is dramatically reconstructed with respect to the normal state, making the utility of the extracted data quite limited (for a review of the QO measurements on 122 parent compounds see, e.g., Ref. [50]). It thus comes as no surprise that the most insightful experiments up to date have been carried out on the extremely overdoped and the end members of the $BaFe_2(As_{1-x}P_x)_2$ series. An exemplary set of experimental data obtained on $BaFe_2P_2$ and a comparison to the corresponding ab initio calculations carried out for the same compound [55] are shown in Fig. 2.6d. The observation of multiple frequency sets in the Fourier-transformed raw data clearly confirms the multiband character of this iron pnictide. The schematic partial Fermi surfaces on the right demonstrate how these frequencies and their angular dependence can be assigned to different hole and electron Fermi pockets of the theoretically predicted band structure. The solid lines in the figure correspond to the angular dependence of the QO signal predicted by the DFT calculation, displaced by $+68/+58$ meV for the inner/outer electron sheets and -113 meV for the inner hole sheet [55]. The band displacements required to reconcile the first-principles results with the experimental data on $BaFe_2P_2$ are quite close to those inferred from the analogous analysis of the data obtained on $SrFe_2P_2$ compounds [59, 60]. It is clear from the comparison in Fig. 2.6d that apart from the band shifts the agreement of experiment and theory is remarkable, implying that the calculation captures the out-of-plane shape of the Fermi surface rather well. Very similar results and conclusions have been obtained for intermediate phosphorus doping levels within the same series [61, 62]. Finally, the renormalization of the quasiparticle mass m^*/m_{band} can be determined from the comparison of the effective quasiparticle mass m^* extracted from the temperature dependence of the QO signal at a particular angle with the effective mass m_{band} predicted by DFT calculations. At terminal phosphorus doping levels this renormalization is of order 2 and largely uniform for all orbitals [60] but gets enhanced to $\sim$4 for certain orbitals upon approaching the superconducting phase [62].

2.4 Magnetic Properties

The proximity of superconductivity to strong antiferromagnetism, observed in both the cuprate and the iron-based superconductors, is believed be another key prerequisite for high-temperature superconductivity [56], along with the two-dimensionality of the Fermi surface discussed above. Although the antiferromagnetic phases are quite different in the cuprates and pnictides, one thing is common: the antiferromagnetic and superconducting order parameters appear to compete in their fully ordered forms but superconductivity is strongest where the antiferromagnetic long-range order has just disappeared completely. This suggests that, although superconductivity is destroyed by long-range antiferromagnetic order, it is in fact driven by the fluctuations of electron spins, which are strongest (but already weak enough not to give rise to the competing long-range order) at the border with the antiferromagnetic phase. A large body of experimental data strongly supports this conclusion: INS revealed a resonance mode, similar to the one observed in the cuprate superconductors, at the value of the antiferromagnetic ordering vector in all iron-based superconductors [8]; the signatures of the mediating boson, glueing Cooper pairs together, revealed with other techniques are also largely consistent with the energy of this resonance mode [21, 23, 63]. This makes the study of the magnetic properties of the iron-based materials indispensable to a complete understanding of their superconducting pairing mechanism.

2.4.1 Antiferromagnetic Ground State

The ground state of the antiferromagnetic phase has been studied in detail in all known iron-based compounds by means of neutron scattering techniques. The literature on the subject is extensive and has already been reviewed, for instance, in Ref. [8]. Therefore, it appears more instructive to review the most relevant to the subsequent discussion characteristic common features of antiferromagnetism in the iron-based materials through the prism of the dichotomy between its local and itinerant character, which so far has received less attention (this discussion is to some extent based on Ref. [64]). To compare these two manifestations of magnetism and to what extent they relate the cuprate and iron-based materials, one first needs to define them. Magnetism is called *local* if the the magnetic state results from the ordering of the already present in the normal state local magnetic moments of the electrons from the outer shell of one of the atomic species comprising the crystal lattice. These moments are usually large because each unpaired electron spin contributes 0.5 μ_B to the total moment. This type of magnetism is most clear in insulating compounds since the absence of delocalized (itinerant) electrons makes all magnetic moments localized. The parent compounds of the cuprate superconductors, being inherent Mott insulators (i.e. their insulating behavior is due to very strong electron–electron correlations [51]; most of these compounds emerge as metals in first-principles calculations, which do not fully account for correlations), constitute a good example of local antiferromagnets. Figure 2.7a shows the diagonal-stripe antiferromagnetic ground state of one of the cuprate parent compounds: La_2CuO_4. It has a sizable magnetic moment of 0.5 μ_B [68], consistent with its local character. The magnetic Bragg peaks formed by this periodic arrangement of the local magnetic moments are located at the $(\pm\pi, \pm\pi)$ points of the first Brillouin zone, shown in Fig. 2.7e (or (1/2,1/2) reciprocal lattice units in the [H,K] notation of the two-dimensional reciprocal space $(2\pi/a, 2\pi/b, 2\pi/c)$ used in the figure). Since the cuprate parent compounds are insulating, whereas superconductivity clearly requires itinerant electrons to exist for them to eventually get bound into delocalized Cooper pairs, doping of either type (electron or hole) is necessary to induce superconductivity in these materials. As a matter of example, and to demonstrate the single-band character of superconductivity in the cuprates, we show the Fermi surface of a slightly doped ($x = 0.07$) but not yet superconducting $La_{1-x}Sr_xCuO_4$. This material shows a single large electronlike Fermi pocket centered at the corners of the Brillouin zone.

In contrast, magnetism is called *itinerant* if the magnetic state results from spontaneous self-organization of the itinerant-electron spin density into a periodic magnetic structure, called a spin-density wave. In this type of magnetism, no magnetic moment needs to be present in the normal state (and it is often not), since the ionic shells and the itinerant-electron spin density can both be spin-compensated. Moreover, the value of the ordered magnetic moment can be much smaller than in the local case because (a) not all itinerant electrons might participate in the formation of this moment and (b) the total number of free electrons can be much smaller in the case of a doped (rather than inherently metallic) material. The aforementioned spontaneous formation of a finite itinerant magnetic moment results from a nesting instability (good geometrical match in the shape) between two sheets of the Fermi surface [69], which is generally strongly enhanced by the low dimensionality of the latter. A classical example of a purely itinerant antiferromagnet is elemental chromium [70].

The situation in the iron-based compounds is significantly more complex. It is currently believed that these compounds are located between purely local magnets like the cuprates and purely itinerant magnets like elemental chromium [64].

Indications of the itinerant character of antiferromagnetism in the iron-pnictides:

- Moderate electronic correlations. The degree of electron–electron correlations can be assessed by directly comparing the band structure obtained in LDA calculations (which neglects most of electronic correlations) with experiment, such as ARPES or QO measurements, as well as by comparing the kinetic energy of itinerant charge carriers approximated by the partial f-sum rule for the optical conductivity [71] in experiment and theory. In both cases one arrives at the same conclusion that the electronic correlations in iron-based materials are moderate and therefore unlikely to create distinct local moments like in the cuprate materials.
- Nesting is good in most compounds. It has been predicted by first-principles calculations and largely confirmed by extensive ARPES measurements that most of iron-based compounds exhibit rather good nesting between hole- and electron-like pockets of the Fermi surface, making them prone to a spin-density-wave

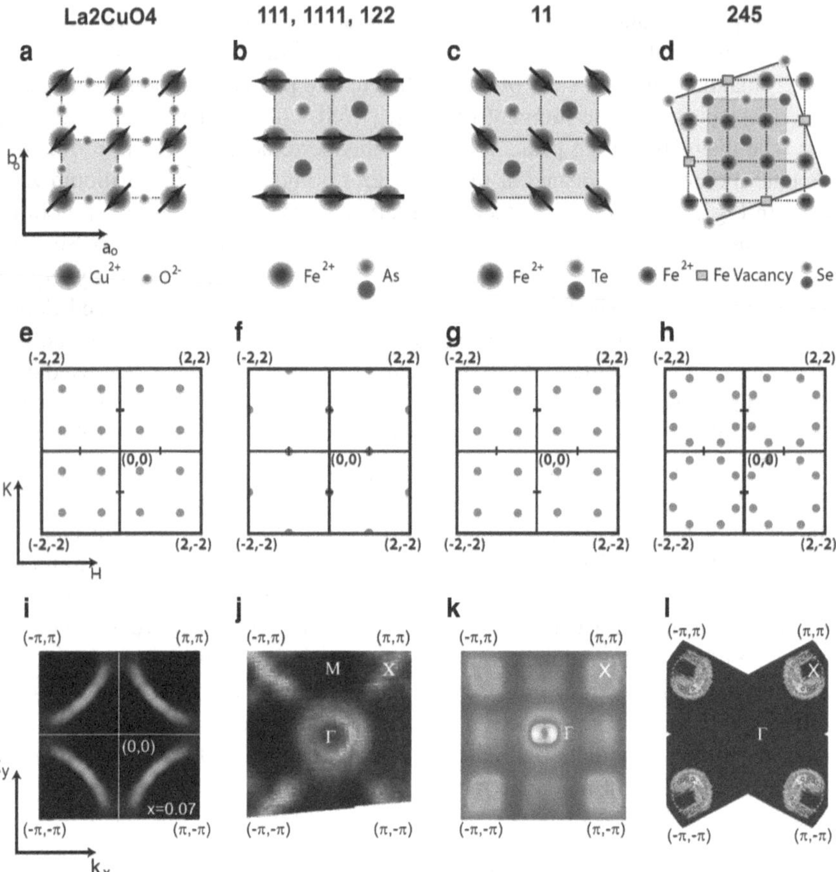

Fig. 2.7 **a–d** Antiferromagnetic ground state of several representative parent compounds of the cuprate and iron-based superconductors: La_2CuO_4 (**a**); LiFeAs (111), LaFeAsO (1111), and $BaFe_2As_2$ (122)—panel (**b**); FeTe (11)—panel (**c**); $K_2Fe_4Se_5$ (245)—panel (**d**). The chemical unit cell (in the case of La_2CuO_4) and the pseudotetragonal unit cell (which approximates the low-temperature orthorhombic unit cell, with the lattice parameters $a_{\mathrm{ort}} \approx b_{\mathrm{ort}} \approx \sqrt{2}a_{\mathrm{tet}}$) are shown in *light green*. **e–h** Magnetic Bragg peaks in the reciprocal space of the respective unit cells marked in *green*. **i–l** Corresponding Fermi surfaces in the normal state (above the magnetic transition temperature) obtained with ARPES: **i** $La_{2-x}Sr_xCuO_4$ at a doping level of $x = 0.07$; **j** $Ba_{1-x}K_xFe_2As_2$ at a doping level of $x = 0.3$; **k** FeTe; **l** $K_2Fe_4Se_5$. In all cases, the Brillouin zone depicted corresponds to the chemical unit cell of the compounds (comprising two Fe atoms (Fig. 2.3c) in the case of iron-based materials), whereas the characteristic points in the reciprocal space (Γ, X, and M) belong to the unfolded Brillouin zone, which corresponds to the unit cell of the iron sublattice comprising one Fe atom (Fig. 2.3b; relevant to magnetic scattering experiments). Panels **a–h** reprinted by permission from Macmillan Publishers Ltd: Nature Physics Ref. [64], copyright (2012). Panels **i**, **k**, and **l** are figures reprinted with permission from Refs. [65–67], respectively. Copyright (2006, 2009, 2011) by the American Physical Society. Panel **j** reprinted by permission from Macmillan Publishers Ltd: Nature Ref. [54], copyright (2009). In panels **i**, **k**, and **l**, the original Fermi surface maps have been symmetrized according to the (pseudo)tetragonal symmetry to cover the entire first Brillouin zone

instability. Figures 2.7j, k show well-nested quasicircular hole and electron sheets of the Fermi surface in four major classes of iron-based materials. In the case of $Ba_{1-x}K_xFe_2As_2$ the high-temperature Fermi-surface reconstruction has already distorted the otherwise circular shape of the outer electron Fermi pocket into 'blades' (Fig. 2.7j and the discussion above).

- Observed ordered antiferromagnetic moment is small in most compounds. In 111, 1111, and 122 compounds the ordered antiferromagnetic moment of 0.3–1.0 μ_B [8] is quite small, bearing in mind that there are six electrons in the Fe-3*d* orbital.
- Spin-wave dispersion is significantly broadened at high energies [72]. The broadening can be ascribed to particle-hole excitations across the Fermi surface, as supported by RPA calculations in an itinerant model [73]. In a stark contrast to this broadening is the instrumental-resolution-limited spin-wave dispersion of La_2CuO_4 [74].

Indications of the local character of antiferromagnetism in the iron-pnictides:

- Observation of different antiferromagnetic ground states in the iron-based compounds although their Fermi surfaces are generally very similar: Figures 2.7j, k show that for 111, 1111, 122, and 11 materials the Fermi surface shows the best nesting along the Γ–X direction in the unfolded magnetic 1-Fe Brillouin zone, which implies the $(\pi, 0)$ ordering vector and the presence of the corresponding Bragg peak in the reciprocal space. Whereas in the 111, 1111, and 122 materials this is indeed the case, as shown in Fig. 2.7f, for 11 compounds the Bragg peaks are located at $(\pm\pi/2, \pm\pi/2)$ points of the magnetic Brillouin zone, in a stark disagreement with the nesting picture.
- Observation of a sizable Fe magnetic moment even at room temperature for most iron-based materials, which is, moreover, comparable to its ordered counterpart. Such observations have been made even for compounds that do not order magnetically at any temperature (e.g. LiFeAs) [64, 75].
- ARPES on LiFeAs shows no nesting but, nevertheless, a sizable bandwidth renormalization of $\sim$3 [64, 76]. On the other hand, LiFePO shows a rather good nesting but no magnetic order [64].

All these considerations strongly indicate that antiferromagnetism in iron-based compounds has a strongly dual character with the signatures of both local and itinerant magnetic moments.

2.4.2 Spin Fluctuations

In superconducting compounds, even in those where superconductivity coexists with antiferromagnetism, all itinerant electrons that do not participate in itinerant magnetism but rather get bound into superconducting Cooper pairs are in a paramagnetic state. Although the ordered magnetic moment is zero in an itinerant paramagnet, statistical fluctuations of electron-spin density may still be significant [77]. Depending on the time scale of these fluctuations in a given material they can be detected by various techniques such as INS, NMR, μSR, etc. In the iron-based superconductors, just like in their cuprate counterparts, such spin fluctuations have indeed been observed and have been shown to occur at the same wave vector in the Brillouin zone as the static magnetism of the respective parent compounds [9, 20]. The properties of spin fluctuations in the 122 iron arsenides in the normal and superconducting state have been extensively discussed in Ref. [15]. The authors have shown that the symmetry and the in-plane anisotropy of these spin fluctuations are well captured by density-functional and itinerant-electron calculations in the random-phase approximation and can be traced back to the structural properties of the host compounds without invoking exotic nematic scenarios with a symmetry-broken electronic state.

2.5 Superconducting Properties

While the mechanism of superconductivity in the iron-based superconductors is under debate, it has been shown theoretically that the electron–phonon coupling in these compounds is very weak and thus unable to account for the observed high superconducting transition temperatures [78] (for an overview of the superconducting gaps and transition temperatures of a large set of iron-based superconductors see Ref. [9]). On the other hand, theoretical treatment has shown that spin fluctuations can, in principle, lead to an *effective* attractive interaction between itinerant electrons and thus to their pairing and the formation of a superconducting condensate [79]. A large body of experimental data obtained on compounds from both the cuprate and the iron-pnictide family of superconductors strongly supports this mechanism of superconductivity [21, 23, 63, 80], making spin fluctuations the most plausible candidate for the so-called 'superconducting pairing glue' or 'mediating boson'. In Sect. 4.2 we will provide evidence supporting this statement based on the analysis of the optical conductivity of $Ba_{0.68}K_{0.32}Fe_2As_2$ in the framework of the Eliashberg theory, the most complete theory of superconductivity to date.

Since the bare electron–electron interaction mediated by spin fluctuations is repulsive [82], no formation of Cooper pairs can occur in the simplest s-wave pairing channel (i.e. with a sign-constant superconducting gap). This is true for any pairing–mediating interaction based on the general form of the gap equation. In the simplest case of the BCS theory of superconductivity the latter reads:

$$\Delta_{\mathbf{k}} = -\frac{1}{2}\sum_{\mathbf{l}} V_{\mathbf{kl}} \frac{\Delta_{\mathbf{l}}}{\sqrt{\xi_{\mathbf{l}}^2 + \Delta_{\mathbf{l}}^2}},$$

where $\Delta_{\mathbf{k}}$ is the momentum-dependent superconducting energy gap, $V_{\mathbf{kl}}$ is the reciprocal-space matrix element of the interaction potential, and $\xi_{\mathbf{k}}$ is the one-electron energy at momentum $\mathbf{k}$ relative to the chemical potential. Given that the potential V of a repulsive interaction is positive, in the case of spin-fluctuation-mediated pairing this relation can obviously only be fulfilled if the gap changes sign in some part of the Brillouin zone. This conclusion is further supported by the observation of a resonance enhancement of the spin-fluctuation spectrum below 2Δ in the superconducting state [9, 19, 20, 22, 83–86] (see also Fig. 2.10). According to the algebraic structure of the magnetic scattering intensity with respect to the coherence factors of the superconducting condensate [36], such a resonance can only occur if $\Delta_{\mathbf{k}}$ changes its sign across the Brillouin zone (although recently it has been demonstrated that a similar peak structure in INS spectra can arise, albeit at a somewhat different characteristic energy, even in the case of a sign-constant superconducting gap [87]). For the multiple-sheet Fermi surface of the iron-based superconductors this implies that the sheets coupled by the repulsive spin-fluctuation-mediated interaction must host superconducting gaps of different sign. If, in addition, the gap on each Fermi surface complies with the tetragonal symmetry of the compound and is sign-constant on the sheets at the center of the Brillouin zone, such a pairing symmetry is called extended s-wave symmetry or $s_\pm$ symmetry, where $\pm$ implies that the gap changes its sign from sheet to sheet. In general, tetragonal symmetry D_{4h} allows for four different symmetries of the superconducting order parameter on a single-sheet Fermi surface in the singlet pairing channel [88]. A schematic representation of these superconducting order parameters in the Brillouin zone, together with the group-theoretical notation, wave function notation, and the corresponding basis functions is given in Fig. 2.8. The situation becomes more complex in a multiband case, giving rise to various 'extended' symmetries of the aforementioned four basic types.

While the determination of the relative sign of the superconducting gap across the Fermi surface is a very difficult task and requires sophisticated phase-sensitive experiments [89–91], its magnitude is readily accessible to a broad variety of experimental techniques, such as ARPES, scanning tunneling spectroscopy (STS), Andreev reflection spectroscopy, optical spectroscopy, and many others. In this regard it is

Group-theoretic notation	A_{1g}	A_{2g}	B_{1g}	B_{2g}
Order parameter basis function	constant	$xy(x^2-y^2)$	x^2-y^2	xy
Wave function name	s-wave	g	$d_{x^2-y^2}$	d_{xy}
Schematic representation of $\Delta(k)$ in B.Z.	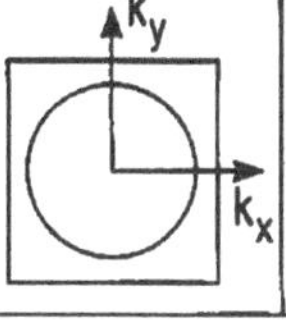 k_y, k_x			

Fig. 2.8 Four symmetries of the superconducting order parameter in the singlet pairing channel allowed by the tetragonal D_{4h} crystallographic symmetry. *Different colors* in the schematic representation of the momentum dependence of the superconducting gap in the Brillouin zone indicate its different signs. Reprinted figure with permission from Ref. [81]. Copyright (2000) by the American Physical Society

important to remember that each technique measures its own *effective* superconducting gap averaged over a certain portion of the Fermi surface, which requires careful extraction of the inherent microscopic gap (when it is possible at all) through modeling of experimental spectra. Since any complex modeling increases the uncertainty of the magnitude of the extracted quantity, it is highly desirable to be able to obtain the value of the superconducting gap with as little data processing as possible. ARPES has the unique capability of not only extracting the value of the superconducting gap with minimal data analysis but also mapping it out for the Fermi surface in the entire Brillouin zone. The knowledge of the complete momentum-dependent superconducting energy gap $\Delta_{\mathbf{k}}$ is indispensable to an adequate theoretical description of superconductivity. As of today, ARPES measurements have already been carried out on most iron-based superconductors [8] and the values of their superconducting gaps have been determined (see Ref. [9] for a digest of the superconducting gaps of the iron-based superconductors obtained with various experimental techniques). In the case of $Ba_{0.6}K_{0.4}Fe_2As_2$, the most complete momentum dependence of the superconducting gap has been given in Refs. [57, 58] and is reproduced in Fig. 2.9. Several characteristic features of the superconducting gap common to all iron-based materials are immediately apparent: first, since superconductivity gaps the Fermi surface in its

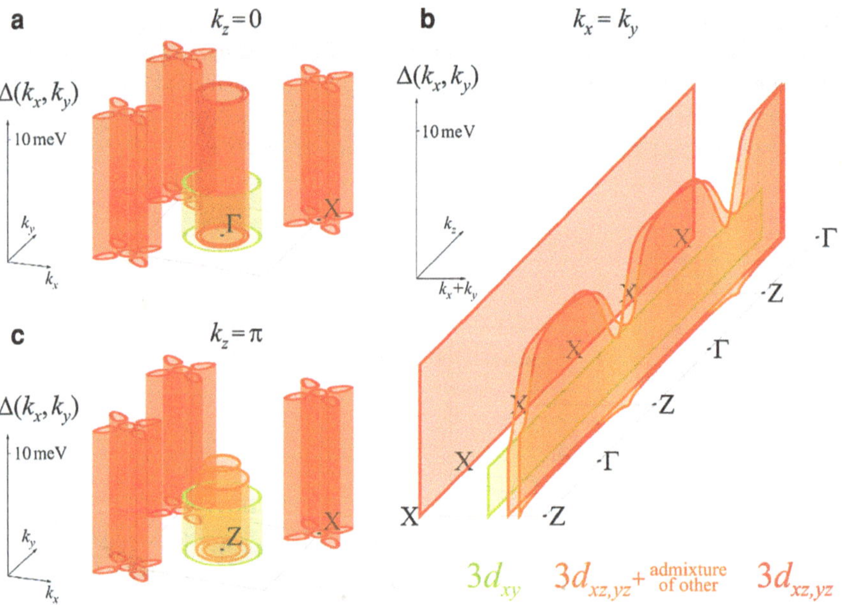

Fig. 2.9 **a** Schematic representation of the momentum-dependent superconducting energy gap of $Ba_{0.6}K_{0.4}Fe_2As_2$ obtained with ARPES at $k_z = 0$. **b** Same for $k_z = \pi$. **c** Momentum dependence of all superconducting gaps in the out-of-plane direction across several Brillouin zones. *Colors* indicate the magnitude of the superconducting gap (increasing from *green* to *red*). Characteristic points in the Brillouin zone refer to the one-iron unit cell of the Fe sublattice. Adapted with author's permission from Ref. [58]

entirety, the existence of multiple electronic bands implies multiple superconducting gaps, and they have indeed been observed; second, the overwhelming majority of the iron-based superconductors exhibit a certain clustering of the magnitudes of the superconducting gaps into two groups. Fig. 2.9a shows how the superconducting gap on the inner two holelike Fermi surfaces at the Γ point and on both Fermi surfaces at the X point of the unfolded 1-Fe Brillouin zone have approximately the same value of 10 meV, whereas on the outer holelike Fermi surface at the Γ point it is significantly smaller, ≈3.5 meV. Although most of the iron-based superconductors, including $Ba_{1-x}K_xFe_2As_2$, exhibit a largely two-dimensional Fermi surface, the dependence of one of the large superconducting gaps on the out-of-plane momentum $\mathbf{k}_z$ shows significant variation, with the magnitude of the gap decreasing by a factor of 3 at $\mathbf{k}_z = \pi$ to match the small gap (see Figs. 2.9b, c). This unexpected out-of-plane dependence of the superconducting gap is at odds with the two-dimensional, cylindrical in the out-of-plane direction, Fermi surface of $Ba_{0.6}K_{0.4}Fe_2As_2$ and strongly suggests that the orbital character of the bands making up the Fermi surface plays a very important role in the mechanism of superconductivity of the iron-based materials. This conclusion is further supported by recent observations of orbital-dependent electronic properties of these compounds [24, 47, 48, 92–94] and several theoretical proposals for an orbital-fluctuation pairing mechanism have been put forward [95, 96].

Although the value of Δ does characterize the energy gap in the quasiparticle excitation spectrum of a superconductor [36], it does not provide direct evidence for the *strength* of the pairing, or how tightly electrons are bound together into Cooper pairs.[3] Determination of the pairing strength is difficult and in most cases involves a significant amount of modeling, although in some cases some insight can be gained already from a simple comparing of the raw data. For instance, by comparing the fingerprints of superconductivity in specific-heat data obtained on high-quality $Ba_{0.68}K_{0.32}Fe_2As_2$ single crystals to those of well-known superconductors, P. Popovich et al. came to the conclusion that this particular compound is a strongly-coupled superconductor [21]. Similar analysis has been applied to $Ba(Fe_{1-x}Co_x)_2As_2$ [97] and other iron-based compounds. The strength of the superconducting pairing interaction can also be determined from the quasiparticle effective-mass renormalization, which can be inferred from QO measurements, as discussed earlier in this chapter. In general, the cyclotron effective-mass renormalization (relevant to the QO technique) of free charge carriers can be defined as $m^*/m_{\text{band}} = 1 + \sum_i \lambda_i$, where λ_i is the coupling constant characterizing the interaction strength of electrons with all other bosonic excitations existing in the system (phonons, spin or orbital fluctuations, excitons, etc.) [98, 99]. One should be very cautious, however, because all interactions, including those that do not lead to superconducting pairing or might even weaken it, contribute to this effective-mass renormalization. Therefore, QO measurements only provide the upper bound on the superconducting pairing strength.

---

[3] Although the superconducting gap does depend on the coupling strength, its dependence on other quantities complicates the extraction of the pairing strength alone [36].

One of the simplest but at the same time the roughest ways to gauge the superconducting coupling strength directly is by computing the ratio $2\Delta/k_B T_c$. While in the single-band case of the BCS theory this gap ratio does not depend on λ and is equal to 3.5, in the multiband case it does acquire a certain dependence on λ [100]. However, already in the two-band case with purely interband interaction the BCS theory gives qualitatively incorrect results at all coupling strengths (even in the weak-coupling limit) due to the complete neglect of the quasiparticle mass renormalization via the pairing interaction. The multiband BCS theory must, therefore, be dismissed in favor of the more complete Eliashberg theory of superconductivity. The latter predicts that the gap ratio grows monotonously with the coupling strength even in the multiband case [101]. This monotonous dependence allows one to estimate the coupling strength by taking a ratio of two easily accessible in experiment quantities: the superconducting energy gap and transition temperature. A comparison of gap ratios for a large number of various superconducting compounds has been carried out in Ref. [9] and revealed that iron pnictides are located between the cuprate and the conventional superconductors with respect to the coupling strength but show significant variation, spanning the range from weak to intermediate coupling.

Both spin fluctuations and Cooper pairs are hosted by the same electronic subsystem, namely, itinerant electrons. Therefore, it comes as no surprise that the development of the superconducting state should have a strong effect on the properties of spin fluctuations, irrespective of the source of the pairing interaction. Such a feedback effect has been observed experimentally in many different families of superconductors and consists in the opening of a spin gap in the spin-fluctuation spectrum, as illustrated in Fig. 2.10 for the case of $BaFe_{1.85}Co_{0.15}As_2$ and $Ba_{0.6}K_{0.4}Fe_2As_2$. Upon entering the superconducting state the low-energy spectrum of spin fluctua-

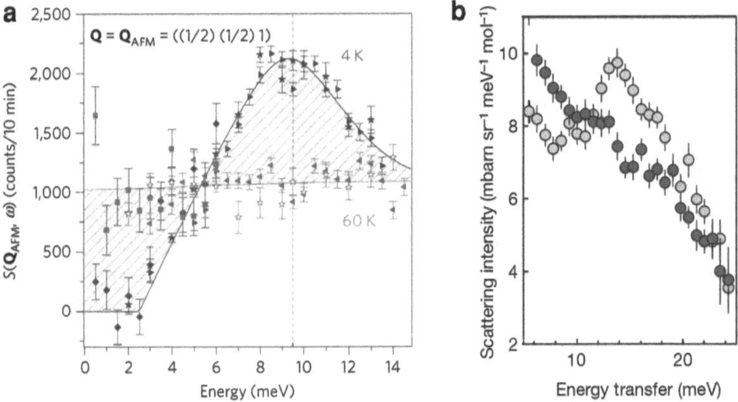

Fig. 2.10 **a** Magnetic scattering function of $BaFe_{1.85}Co_{0.15}As_2$ in the normal and superconducting state at 60 and 4 K, respectively, after a background correction. Reprinted by permission from Macmillan Publishers Ltd: Nature Physics Ref. [20], copyright (2009). **b** Neutron scattering intensity (cross-section) from $Ba_{0.6}K_{0.4}Fe_2As_2$ in the normal and superconducting state at 50 and 7 K, respectively. Reprinted by permission from Macmillan Publishers Ltd: Nature Ref. [19], copyright (2008)

tions shows dramatic enhancement below 2Δ at a characteristic frequency $\omega_{\rm res}$, with a concomitant suppression at even lower frequencies. Due to the conservation of the total electron spin in the system, the total spectral weight (the area under the curve) of the magnetic scattering function $S(\mathbf{Q}, \omega)$[4] does not change between the normal and the superconducting state. A peak in the spin-fluctuation spectrum only occurs if the coherence factors of the magnetic dipole interaction interfere constructively, which requires that the superconducting gap change its sign across the Fermi surface, as mentioned above. The above discussion holds true for any type of pairing interaction. If, however, superconducting pairing is mediated by spin fluctuations themselves, a successful theory of superconductivity would have to account for this feedback effect self-consistently. Unfortunately, it introduces a significant degree of additional complexity into the problem and makes its handling challenging. If only a qualitative description of superconducting properties is sought, then, to the first approximation, the feedback effect of superconductivity on spin fluctuations can be neglected, as it will be done in the analysis of the far-infrared conductivity of $Ba_{0.68}K_{0.32}Fe_2As_2$ in the superconducting state in Sect. 4.2. Some of the consequences of the feedback effect for the optical conductivity of superconductors have been addressed in Ref. [102].

2.6 Properties of the $A_2Fe_4Se_5$-Type Compounds (A = K, Rb, Cs)

After the discovery of the 122 iron pnictides [5], massive research effort has been made to identify and characterize other compounds of this family. Numerous elements of the periodic table have been used for doping or substitution and as the base elements of parent compounds [8]. In the former case, both aliovalent substitution resulting in doping with holes [$Ba_{1-x}K_xFe_2As_2$, $Ba_{1-x}Na_xFe_2As_2$] and electrons [$Ba(Fe_{1-x}Co_x)_2As_2$, $Ba(Fe_{1-x}Ni_x)_2As_2$] and isovalent substitution changing the electronic structure without affecting the position of the Fermi level [$Ba(Fe_{1-x}Ru_x)_2As_2$, $BaFe_2(As_{1-x}P_x)_2$] have been explored. In the search of new parent compounds several chemical elements have been considered as intercalants: Ca, Sr, and Ba have all produced compounds with quite similar gross physical properties although certain subtle changes have been discovered upon more detailed scrutiny [8]. In all cases the synthesis of single-phase materials throughout the entire phase diagram has been achieved, with superconductivity and antiferromagnetism coexisting on a microscopic level in its certain regions [28]. Until late 2010 only Fe–As-based compounds had existed in the 122 class of the iron-based materials, when at the very end of the year a new type of the 122 compounds based on Fe–Se

---

[4] The magnetic scattering function is connected to the differential neutron scattering cross-section (intensity) per atom via $d^2\sigma/Nd\Omega dE_{\rm f} = (k_{\rm f}/k_{\rm i})\,|b|^2\,S(\mathbf{Q}, \omega)$, where N is the total number of atoms, Ω is the scattering solid angle, $E_{\rm f}$ is the final energy, $k_{\rm f,i}$ is the final (initial) wave vector of neutrons, and b is the scattering length.

layers was discovered and reported to have a superconducting transition temperature as high as 32 K [103–108]. Three members of this $A_x Fe_{2-y} Se_2$ class, differing only in the intercalating atom (A = K, Rb, and Cs), have been successfully synthesized and extensively studied. These compounds were conceived as an extension of the 122 class of the iron arsenides and were expected to share most of the common characteristics of the latter. However, very soon it has been realized that, although the crystallographic structure of the 122 iron arsenides and chalcogenides was in many respects alike, many physical and chemical properties have been found to differ drastically. Even though the new superconductors displayed a rather high superconducting transition temperature, comparable to the optimally 122 iron arsenides, in the former this transition temperature was achieved *without* any doping or substitution and turned out to be independent of the intercalating alkaline metal [103–105, 108]. The microstructural properties of the Fe–Se layer were also found to remain largely unaffected by the choice of the intercalating atom [109]. Further, unlike the single-phase 122 iron arsenides, these new compounds have defied all efforts to synthesize a bulk single-phase material. Granted that this could be taken as the indication of the relatively low quality of single crystals synthesized to date, there is significant experimental evidence that phase separation in the 122 iron chalcogenides is essentially inherent in its nature, as will be discussed at the end of this section and in Sect. 4.4.

Subsequent extensive bulk-sensitive measurements of the 122 iron chalcogenides revealed that, although the superconducting *shielding* volume fraction inferred from ac susceptibility measurements in the best-quality compounds approached 100 % [22, 110] (100 % diamagnetic shielding implies $4\pi\chi = 1$ in the zero-field-cool configuration), its actual fraction made up only 10–20 % of the sample volume [111–114], immediately implying that superconductivity in these materials bears a percolating character. This conclusion was later confirmed in a study of the dependence of the morphology of the superconducting-phase network in potassium-doped compounds on crystal growth parameters and post-processing [115]. The existence of phase separation in the 122 iron chalcogenides has now been observed by practically every experimental technique [27, 111–114, 116–130] and is thus reliably established. Quite interestingly, phase separation in the out-of-plane direction was discovered to occur on the nanoscale in the shape of quasiregularly alternating layers of the superconducting and antiferromagnetic phase [114].

Given that the superconducting phase occupies a very small fraction of the sample volume, some of its properties remained elusive for almost 2 years of intensive research. Therefore, we first review the main features of the antiferromagnetic majority phase, which have all been established robustly soon after the discovery of these compounds. Already the first structural refinement of single-crystal X-ray diffraction data obtained on these compounds has revealed that, unlike the tetragonal I4/mmm symmetry of $ThCr_2Si_2$ type found in all 122 iron arsenides at room temperature, an inherent iron-deficiency order with a chiral $\sqrt{5} \times \sqrt{5} \times 1$ superstructure, present in these materials, reduces the symmetry to I4/m and makes it more appropriate to classify them into the 245 stoichiometry [7] (e.g. $Rb_2Fe_4Se_5$), with the lattice parameters being $a = 8.7653(2)$ Å, $c = 13.8811(5)$ Å. The in-plane structure, including iron vacancies and the $\sqrt{5} \times \sqrt{5}$-reconstructed unit cell (black rectangle), is shown in Fig. 2.1b. The out-of-plane crystallographic structure is identical to the 122 iron arsenides (see Fig. 2.1a). Such a $\sqrt{5} \times \sqrt{5} \times 1$ iron-vacancy–ordered structure implies, in the absence of additional iron defects, that 1/5 of all iron ions are missing in comparison to the 122 system. The experimentally observed C4 rotational symmetry of I4/m further limits the possibility of alkali-metal vacancies to either none (all sites occupied) or one alkali-metal ion missing (1/5 of all alkali-metal ions) at the center of the simple tetragonal 122 unit cell, shown in Fig. 2.1a. These two considerations imply that the only possible chemical compositions of the antiferromagnetic phase are $A_xFe_{1.6}Se_2$, with x equal to either 1 or 0.8. Recent STM/STS measurements on thin films of $K_2Fe_4Se_5$ have confirmed the absence of Fe and Se defects in the antiferromagnetic phase [124]. This study also found a slightly different intensity at every fifth potassium site, which would suggest the $x = 0.8$ composition but this evidence remained inconclusive due to possible masking electronic effects. The $x = 0.8$ composition also seems to be consistent with the ratio of ^{37}Rb NMR intensities in the superconducting and antiferromagnetic phase [113].

Early neutron diffraction measurements established that the 122 iron chalcogenides undergo almost simultaneous structural and magnetic-ordering transitions at a very high temperature on the order of 550 K [133]. It was also discovered that these compounds possess a magnetic moment on iron atoms of about 3.3 μ_B [133], which is unusually large for the iron pnictides. Moreover, this ordered antiferromagnetic moment was found to be aligned along the crystallographic c-axis, in contrast to its in-plane orientation in all other magnetically ordered iron-based materials [8], and susceptible to superconductivity, which indicates a certain degree of interplay between these two phases [133]. In the same neutron diffraction study an unusual antiferromagnetic ground state was revealed, which featured clustering of iron ions into plaquettes of four, ordered ferromagnetically within the plaquettes and antiferromagnetically between them, as shown in Fig. 2.11a. This exotic ground state was quickly confirmed by the first-principles calculations of the electronic and magnetic structure of the 245 materials [131]. These ab initio calculations also predicted that the antiferromagnetically ordered phase is a semiconductor with a direct band gap of $\approx$0.5 eV. In subsequent measurements of the optical conductivity of $Rb_2Fe_4Se_5$ a direct band gap was indeed identified [119], with a value of 0.45 eV, very close to the predicted value. The semiconducting and metallic optical properties of $Rb_2Fe_4Se_5$ will be considered in more detail in Sect. 4.3.

The properties of the minority metallic phase, unlike those of the antiferromagnetic phase, have seen much debate. Even the chemical composition of the metallic phase long remained a mystery. Opinions ranged from the stoichiometric 122 composition identical to that of the 122 iron arsenides (see e.g. Ref. [124]) to the chemical composition of the antiferromagnetic phase but without iron-vacancy order [134]. It is worth noting that while the latter implies rather moderate doping levels (depending on the relative concentrations of the alkaline metal and Fe with respect to the fully compensated 245 composition), the former case would result in a doping level of 0.5 electron per Fe lattice site (including the iron-vacancy sites), which corresponds to an extremely overdoped case of the 122 iron arsenides, beyond the superconducting dome in Fig. 2.4 for $Ba(Fe_{1-x}Co_x)_2As_2$. This debate has now largely been settled by a combination of an ARPES study [67], theoretical RPA calculations [135], INS [84, 85], and NMR measurements [113], which have provided compelling evidence for the $A_{0.3}Fe_2Se_2$ composition of the superconducting phase. It corresponds to a doping level of 0.15 electrons per Fe atom, much closer to the optimal doping level of $Ba(Fe_{1-x}Co_x)_2As_2$ (a doping level of 0.15 electrons per Fe atom corresponds to $x = 0.3$ on the phase diagram of $Ba(Fe_{1-x}Co_x)_2As_2$ in Fig. 2.4). The same $A_{0.3}Fe_2Se_2$ composition of the metallic phase seems to stabilize in the $A_xFe_{2-y}Se_2$ iron chalcogenides regardless of the type of the intercalating alkaline metal (A = K, Rb, and Cs) [85]. We would also like to point out that while the first ARPES measurements on a $K_2Fe_4Se_5$ superconductor, mentioned above, discovered the absence of hole pockets at the center of the Brillouin zone (and, in fact, the absence of any holelike Fermi-surface sheets whatsoever) at $k_z = 0$, as shown in Fig. 2.7l, subsequent measurements at $k_z = \pi$ revealed an additional electronlike pocket at the Z point of the Brillouin zone, illustrated schematically in Fig. 2.11b. The latter study also discovered nodeless superconducting gaps on all Fermi surfaces, with the central pocket developing a gap of 5 meV and the corner pockets a gap of 10 meV (see Fig. 2.11c). Recent STM/STS measurements on thin films of the same material have found twice smaller gaps [124].

Phase separation in the 122 iron chalcogenides makes the extensive investigation of this system with the entire arsenal of experimental condensed-matter research challenging, most prominently due to the intrinsic volume-averaging character of many experimental techniques, such as optics. In the latter case, useful information about the properties of the superconducting phase can only be extracted by resorting to an effective-medium approximation [121, 122], which, however, requires the detailed knowledge of the geometry and volume fraction of the minority phase. Fortunately,

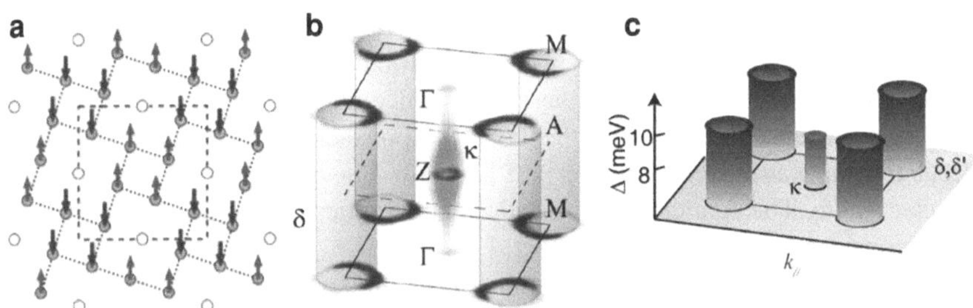

Fig. 2.11 **a** Block-checkerboard-type antiferromagnetic ground state of the $A_xFe_{2-y}Se_2$ compounds identified in first-principles magnetic calculations based on experimentally obtained crystallographic structure. This magnetic pattern comprises clusters of four ferromagnetically coupled iron magnetic moments, with the clusters coupled antiferromagnetically between themselves. **b** Schematic representation of the complete Fermi surface of the 122 iron-chalcogenide superconductors obtained with ARPES. Note the absence of any Fermi pockets at the Γ point of the 2-Fe Brillouin zone and a small electronlike Fermi surface sheet at the Z point. **c** Superconducting gaps observed with ARPES at $k_z = \pi$ on the Fermi surfaces denoted as δ and κ in **b**. Panels **a**, **b**, and **c** are figures reprinted with permission from Refs. [131], [116], and [132], respectively. Copyright (2011, 2012) by the American Physical Society

many imaging and scanning techniques (STM/STS, ARPES, s-SNOM) possess a sufficient spatial resolution to be able to probe the superconducting and antiferromagnetic phases separately, thus providing reliable information about their properties. First ARPES measurements carried out on a $K_2Fe_4Se_5$ compound revealed sizable electron pockets of the Fermi surface at the corners of the 2-Fe Brillouin zone but observed no holelike pockets at its Γ point, as shown in Fig. 2.71, thus excluding the likely realized in most other iron-based superconductors extended s-wave ($s_\pm$) symmetry of the superconducting order parameter originating from spin-fluctuation–induced pairing enhanced by the nesting of holelike pockets at the center and electronlike pockets at the corners of the Brillouin zone. In the absence of holelike Fermi pockets at the center of the Brillouin zone one can still achieve spin-fluctuation–mediated superconductivity under the assumption of a $d_{x^2-y^2}$ superconducting order parameter in the unfolded 1-Fe Brillouin zone [135–137]. Since there are no Fermi pockets along the nodal lines of this type of the order parameter, the resulting superconducting state would be fully gapped (without nodes) in the unfolded Brillouin zone. However, one should bear in mind that the unfolded Brillouin zone, while quite useful in the analysis of magnetic properties, is a convenient abstraction and in reality the body-centered tetragonal symmetry of the iron-based superconductors leads to its folding along the $\mathbf{Q} = (\pi, \pi, \pi)$ vector of the reciprocal space, as can be inferred from Fig. 2.3. Given a fully gapped d-wave state in the unfolded Brillouin zone, the aforementioned folding would lead to an overlap and hybridization of the Fermi surfaces with the '+' and '−' sign of superconducting order parameter, which would obviously result in the formation of nodes [136], as shown in Fig. 2.12. As a consequence, the same (hybridized) Fermi surface would exhibit both positive and negative values of the superconducting gap. These considerations appear to be supported by recent INS measurements combined with calculations in the random-phase approximation [84], which explained the unusual quadruple structure of the neutron resonance peak in the superconducting state shown in Fig. 2.12 in the itinerant picture as a result of enhanced quasiparticle interaction between the well-nested portions of the Fermi pockets with the opposite signs of the superconducting gap (Fig. 2.12). However, this picture has been challenged by a recent ARPES study carried out at the out-of-plane wave vector $k_z = \pi$ [132], already mentioned above, which revealed nodeless superconducting gaps on all electron Fermi surfaces, including the small electron pocket at the Z point of the 2-Fe Brillouin zone (Figs. 2.11b, c). If the symmetry of the superconducting order parameter were indeed d-wave in the unfolded Brillouin zone (with its hybridized counterpart in the folded Brillouin zone), this small electron pocket would necessarily exhibit nodes by symmetry (see Fig. 2.12a).

As a matter of conclusion, we would like to comment on the issue of whether the separation of the superconducting and antiferromagnetic phases is extrinsic (thus allowing for the superconducting phase to be isolated) or intrinsic (implying that the superconducting phase only exists in the proximity of the antiferromagnetic phase and not by itself). Although the practice of experimental condensed-matter research shows that macroscopic phase separation is often an indication of the low quality of single crystals in view, in the case of the 122 iron chalcogenides, significant experimental evidence points towards the inherent character of the coexistence of

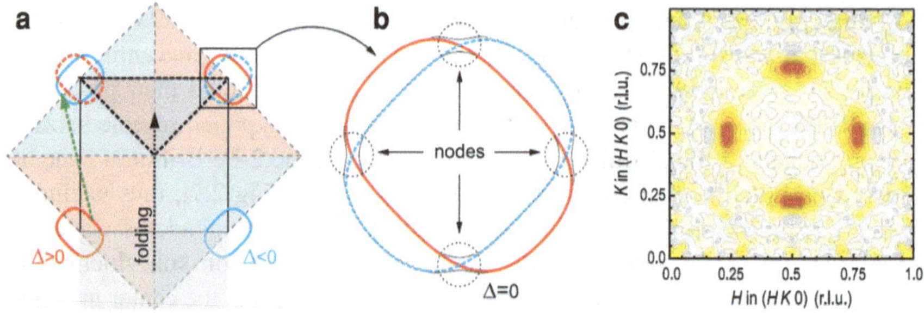

Fig. 2.12 **a** Fermi surface and the schematic structure of the superconducting order parameter in the $A_xFe_{2-y}Se_2$ superconductors. The spin-fluctuation–mediated interaction leads to a $d_{x^2-y^2}$-wave symmetry of the superconducting gap in the unfolded Brillouin zone (*dashed square*) [135, 136], illustrated here as a positive (*red Fermi surface*) and negative (*blue Fermi surface*) superconducting gap in the respective quadrants of the Brillouin zone. Since the $(\pm\pi, \pm\pi)$ nodal lines (*dashed lines* connecting the corners of the unfolded Brillouin zone) do not cross any of the Fermi surfaces, the superconducting gap exhibits no nodes. However, when the full body-centered tetragonal symmetry is taken into account, the hypothetical unfolded Brillouin zone is folded along the $\mathbf{Q} = (\pi, \pi, \pi)$ vector, whose in-plane projection is shown here as a *black dotted arrow*. The smaller, folded Brillouin zone (*black square*) is formed by making equivalent all parts of the unfolded Brillouin zone connected by the folding vector, i.e. the *gray shaded area* is transferred by $\mathbf{Q}$ to become the portion of the folded Brillouin zone marked with *thick dashed lines*. This process results in now two electronlike Fermi surfaces at the corners of the new Brillouin zone but obviously with different signs of the superconducting order parameter. **b** Result of the hybridization of the two electronlike pockets at each corner of the folded Brillouin zone. In the hybridization areas the superconducting gap must change its sign while remaining on the same Fermi surface and therefore must go through zero. **c** Neutron resonance mode detected in the superconducting state of $Rb_2Fe_4Se_5$. Reprinted figure with permission from Ref. [84]. Copyright (2012) by the American Physical Society. Similar features have been observed in other compounds of the $A_xFe_{2-y}Se_2$ family (A = K, Rb, and Cs) in Refs. [22, 85, 86]. The unusual positions of the resonance peaks in the unfolded Brillouin zone can be shown to result from enhanced quasiparticle interaction between the well-nested portions of the Fermi surface (*green dashed arrow* in **a**) [84]

these two phases. Mössbauer measurements under pressure, carried out on single crystals of $Rb_2Fe_4Se_5$, have revealed that superconductivity and antiferromagnetism disappear *simultaneously* with increasing hydrostatic pressure [27]. Such a correlation would be quite unusual for two independent, macroscopically separated phases. In addition, this study observed only a negative pressure effect on the transition temperature, i.e. the latter only decreases with increasing pressure, which indicates that superconductivity in these chalcogenide materials is already 'optimized' to the top of the superconducting dome. A recent neutron diffraction study of a closely related $TlFe_{1.6}Se_2$ material showed that the majority phase acquires a completely different ground state in the presence of phase separation with a minority phase with disordered iron ions [138]. Such a pronounced coupling between these phases appears natural in view of their nanoscale separation in the out-of-plane direction [114], allowing them to interact via lattice strain. The authors of Ref. [138] further infer from their scanning transmission electron microscopy measurements that the *c*-axis lattice parameter of the minority disordered phase is 2 % larger than that of the ordered phase, while the corresponding difference in the in-plane counterpart is less than 0.3 %. Bearing in mind the strong effect of the pnictogen/chalcogen height on the superconducting transition temperature, it appears possible that the role of the presence of the encapsulating antiferromagnetic phase in the superconductivity of the 122 iron chalcogenides is to optimize the crystallographic structure (chalcogen height) by creating effective internal pressure (changing the c-axis lattice parameter) and thus to maximize the superconducting transition temperature. This picture is supported by recent first-principles calculations [139], which revealed a strong influence of the Wyckoff lattice position of the alkaline-metal ion on the electronic properties of the 122 iron chalcogenides and suggested a compelling explanation for re-entrant superconductivity with an even higher superconducting transition temperature upon application of external pressure observed in these compounds [27]. If the role of internal pressure generated on the superconducting phase by the coexisting antiferromagnetic phase is indeed at the root of superconductivity in these materials, it would be impossible to isolate the superconducting phase in free space without applying external pressure. In fact, it means that perhaps the true chemical equivalent of the superconducting phase found in the $A_xFe_{2-y}Se_2$ compounds has already been synthesized but remained completely overlooked because it showed no superconductivity due to the absence of the aforementioned pressure-induced optimization of the c-axis lattice parameter.

References

1. Kamihara, Y., et al. (2006). Iron-based layered superconductor: LaOFeP. *Journal of the American Chemical Society, 128*, 10012–10013.
2. Kamihara, Y., Watanabe, T., Hirano, M., & Hosono, H. (2008). Iron-based layered superconductor $La[O_{1-x}F_x]FeAs(x = 0.05 - 0.12)$ with $T_c = 26$ K. *Journal of the American Chemical Society, 130*, 3296.
3. Nagamatsu, J., Nakagawa, N., Muranaka, T., Zenitani, Y., & Akimitsu, J. (2001). Superconductivity at 39 K in magnesium diboride. *Nature, 410*, 63–64.
4. Zhi-An, R., et al. (2008). Superconductivity at 55 K in iron-based F-doped layered quaternary compound $Sm[O_{1-x}F_x]FeAs$. *Chinese Physics Letters, 25*, 2215.
5. Rotter, M., Tegel, M., & Johrendt, D. (2008). Superconductivity at 38 K in the iron arsenide $(Ba_{1-x}K_x)Fe_2As_2$. *Physical Review Letters, 101*, 107006.
6. Chu, C. W. (2009). High-temperature superconductivity: alive and kicking. *Nature Physics, 5*, 787–789.
7. Bacsa, J., et al. (2011). Cation vacancy order in the $K_{0.8+x}Fe_{1.6-y}Se_2$ system: five-fold cell expansion accommodates 20 % tetrahedral vacancies. *Chemical Science, 2*, 1054.
8. Johnston, D. C. (2010). The puzzle of high temperature superconductivity in layered iron pnictides and chalcogenides. *Advances in Physics, 59*, 803.
9. Inosov, D. S., et al. (2011). Crossover from weak to strong pairing in unconventional superconductors. *Physical Review B, 83*, 214520.
10. Rotter, M., et al. (2008). Spin-density-wave anomaly at 140 K in the ternary iron arsenide $BaFe_2As_2$. *Physical Review B, 78*, 020503.
11. Song, C.-L., et al. (2012). Suppression of superconductivity by twin boundaries in FeSe. *Physical Review Letters, 109*, 137004.

12. Kimber, S. A. J., et al. (2009). Similarities between structural distortions under pressure and chemical doping in superconducting $BaFe_2As_2$. *Nature Materials, 8*, 471–475.
13. Guo, Y., et al. (2012). Continuous critical temperature enhancement with gradual hydrogen doping in $LaFeAsO_{0.85}H_x$ ($x = 0 - 0.85$). *Physical Review B, 86*, 054523.
14. Kittel, C. (2004). *Introduction to solid state physics*. New York: Wiley.
15. Park, J. T., et al. (2010). Symmetry of spin excitation spectra in the tetragonal paramagnetic and superconducting phases of 122-ferropnictides. *Physical Review B, 82*, 134503.
16. Marcinkova, A., et al. (2010). Superconductivity in $NdFe_{1-x}Co_xAsO$ ($0.05 \leq x \leq 0.20$) and rare-earth magnetic ordering in NdCoAsO. *Physical Review B, 81*, 064511.
17. McGuire, M. A., Sefat, A. S., Sales, B. C., & Mandrus, D. (2010). Iron substitution in NdCoAsO: crystal structure and magnetic phase diagram. *Physical Review B, 82*, 092404.
18. Li, Y. K., et al. (2012). Magnetic phase diagram in the Co-rich side of the $LCo_{1-x}Fe_xAsO$ ($L =$ La, Sm) system. *Physical Review B, 86*, 104408.
19. Christianson, A. D., et al. (2008). Unconventional superconductivity in $Ba_{0.6}K_{0.4}Fe_2As_2$ from inelastic neutron scattering. *Nature, 456*, 930.
20. Inosov, D. S., et al. (2009). Normal-state spin dynamics and temperature-dependent spin-resonance energy in optimally doped $BaFe_{1.85}Co_{0.15}As_2$. *Nature Physics, 6*, 178.
21. Popovich, P., et al. (2010). Specific heat measurements of $Ba_{0.68}K_{0.32}Fe_2As_2$ single crystals: evidence for a multiband strong-coupling superconducting state. *Physical Review Letters, 105*, 027003.
22. Park, J. T., et al. (2011). Magnetic resonant mode in the low-energy spin-excitation spectrum of superconducting $Rb_2Fe_4Se_5$ single crystals. *Physical Review Letters, 107*, 177005.
23. Charnukha, A., et al. (2011). Eliashberg approach to infrared anomalies induced by the superconducting state of $Ba_{0.68}K_{0.32}Fe_2As_2$ single crystals. *Physical Review B, 84*, 174511.
24. Lee, G., et al. (2012). Orbital selective Fermi surface shifts and mechanism of high T_c superconductivity in correlated AFeAs ($A =$Li, Na). *Physical Review Letters, 109*, 177001.
25. Gati, E., et al. (2012). Hydrostatic-pressure tuning of magnetic, nonmagnetic, and superconducting states in annealed $Ca(Fe_{1-x}Co_x)_2As_2$. *Physical Review B, 86*, 220511.
26. Stewart, G. R. (2011). Superconductivity in iron compounds. *Reviews of Modern Physics, 83*, 1589–1652.
27. Ksenofontov, V., et al. (2012). Superconductivity and magnetism in $Rb_{0.8}Fe_{1.6}Se_2$ under pressure. *Physical Review B, 85*, 214519.
28. Marsik, P., et al. (2010). Coexistence and competition of magnetism and superconductivity on the nanometer scale in underdoped $BaFe_{1.89}Co_{0.11}As_2$. *Physical Review Letters, 105*, 057001.
29. Park, J. T., et al. (2009). Electronic phase separation in the slightly underdoped iron pnictide superconductor $(Ba_{1-x}K_x)Fe_2As_2$. *Physical Review Letters, 102*, 117006.
30. Li, Z., et al. (2012). Microscopic coexistence of antiferromagnetic order and superconductivity in $Ba_{0.77}K_{0.23}Fe_2As_2$. *Physical Review B, 86*, 180501.
31. Avci, S., et al. (2012). Phase diagram of $Ba_{1-x}K_xFe_2As_2$. *Physical Review B, 85*, 184507.
32. Ma, L., et al. (2012). Microscopic coexistence of superconductivity and antiferromagnetism in underdoped $Ba(Fe_{1-x}Ru_x)_2As_2$. *Physical Review Letters, 109*, 197002.
33. Hashimoto, K., et al. (2012). A sharp peak of the zero-temperature penetration depth at optimal composition in $BaFe_2(As_{1-x}P_x)_2$. *Science, 336*, 1554–1557.
34. Rotter, M., Pangerl, M., Tegel, M., & Johrendt, D. (2008). Superconductivity and crystal structures of $(Ba_{1-x}K_x)Fe_2As_2$ ($x = 0 - 1$). *Angewandte Chemie International Edition, 47*, 7949–7952.
35. Chu, J.-H., Analytis, J. G., Kucharczyk, C., & Fisher, I. R. (2009). Determination of the phase diagram of the electron-doped superconductor $Ba(Fe_{1-x}Co_x)_2As_2$. *Physical Review B, 79*, 014506.
36. Tinkham, M. (1995). *Introduction to superconductivity*. New York: McGraw-Hill.
37. Goldman, A. I., et al. (2008). Lattice and magnetic instabilities in $CaFe_2As_2$: a single-crystal neutron diffraction study. *Physical Review B, 78*, 100506.
38. Ni, N., et al. (2008). First-order structural phase transition in $CaFe_2As_2$. *Physical Review B, 78*, 014523.

39. Yan, J.-Q., et al. (2008). Structural transition and anisotropic properties of single-crystalline $SrFe_2As_2$. *Physical Review B, 78*, 024516.
40. Loudon, J. C., Bowell, C. J., Gillett, J., Sebastian, S. E., & Midgley, P. A. (2010). Determination of the nature of the tetragonal to orthorhombic phase transition in $SrFe_2As_2$ by measurement of the local order parameter. *Physical Review B, 81*, 214111.
41. Blomberg, E. C., et al. (2011). In-plane anisotropy of electrical resistivity in strain-detwinned $SrFe_2As_2$. *Physical Review B, 83*, 134505.
42. Chu, J.-H., et al. (2010). In-plane resistivity anisotropy in an underdoped iron arsenide superconductor. *Science, 329*, 824–826.
43. Tanatar, M. A., et al. (2010). Uniaxial-strain mechanical detwinning of $CaFe_2As_2$ and $BaFe_2As_2$ crystals: optical and transport study. *Physical Review B, 81*, 184508.
44. Chu, J.-H., Kuo, H.-H., Analytis, J. G., & Fisher, I. R. (2012). Divergent nematic susceptibility in an iron arsenide superconductor. *Science, 337*, 710–712.
45. Kasahara, S., et al. (2012). Electronic nematicity above the structural and superconducting transition in $BaFe_2(As_{1-x}P_x)_2$. *Nature, 486*, 382–385.
46. Fernandes, R. M., Abrahams, E., & Schmalian, J. (2011). Anisotropic in-plane resistivity in the nematic phase of the iron pnictides. *Physical Review Letters, 107*, 217002.
47. Fernandes, R. M., Chubukov, A. V., Knolle, J., Eremin, I., & Schmalian, J. (2012). Preemptive nematic order, pseudogap, and orbital order in the iron pnictides. *Physical Review B, 85*, 024534.
48. Yi, M., et al. (2011). Symmetry-breaking orbital anisotropy observed for detwinned $Ba(Fe_{1-x}Co_x)_2As_2$ above the spin density wave transition. *Proceedings of the National Academy of Sciences, 108*, 6878–6883.
49. Damascelli, A., Hussain, Z., & Shen, Z.-X. (2003). Angle-resolved photoemission studies of the cuprate superconductors. *Reviews of Modern Physics, 75*, 473–541.
50. Sebastian, S. E. (2012). Quantum oscillations in iron pnictide superconductors. In N.L. Wang, H. Hosono, P. C. Dai (Eds.), *Iron-based superconductors—materials, properties and mechanisms*. Singapore: Pan Stanford Publishing.
51. Imada, M., Fujimori, A., & Tokura, Y. (1998). Metal-insulator transitions. *Reviews of Modern Physics, 70*, 1039–1263.
52. Charnukha, A., et al. (2011). Superconductivity-induced optical anomaly in an iron arsenide. *Nature Communications, 2*, 219.
53. Mazin, I. I., & Schmalian, J. (2009). Pairing symmetry and pairing state in ferropnictides: theoretical overview. *Physica C, 469*, 614–627.
54. Zabolotnyy, V. B., et al. (2009). (π, π) electronic order in iron arsenide superconductors. *Nature, 457*, 569–572.
55. Arnold, B. J., et al. (2011). Nesting of electron and hole Fermi surfaces in nonsuperconducting $BaFe_2P_2$. *Physical Review B, 83*, 220504.
56. Basov, D. N., & Chubukov, A. V. (2011). Manifesto for a higher T_c. *Nature Physics, 7*, 272–276.
57. Evtushinsky, D. V., et al. (2009). Momentum dependence of the superconducting gap in $Ba_{1-x}K_xFe_2As_2$. *Physical Review B, 79*, 054517.
58. Evtushinsky, D.V., et al. (2012). Strong pairing at iron $3d_{xz,yz}$ orbitals in hole-doped $BaFe_2As_2$. arXiv:1204.2432 (unpublished).
59. Analytis, J. G., et al. (2009). Fermi surface of $SrFe_2P_2$ determined by the de Haas-van Alphen effect. *Physical Review Letters, 103*, 076401.
60. Carrington, A. (2011). Quantum oscillation studies of the Fermi surface of iron-pnictide superconductors. *Reports on Progress in Physics, 74*, 124507.
61. Analytis, J. G., Chu, J.-H., McDonald, R. D., Riggs, S. C., & Fisher, I. R. (2010). Enhanced Fermi-surface nesting in superconducting $BaFe_2(As_{1-x}P_x)_2$ revealed by de Haas-van Alphen effect. *Physical Review Letters, 105*, 207004.
62. Shishido, H., et al. (2010). Evolution of the Fermi surface of $BaFe_2(As_{1-x}P_x)_2$ on entering the superconducting dome. *Physical Review Letters, 104*, 057008.

63. Shan, L., et al. (2012). Evidence of a spin resonance mode in the iron-based superconductor $Ba_{0.6}K_{0.4}Fe_2As_2$ from scanning tunneling spectroscopy. *Physical Review Letters, 108*, 227002.
64. Dai, P., Hu, J., & Dagotto, E. (2012). Magnetism and its microscopic origin in iron-based high-temperature superconductors. *Nature Physics, 8*, 709–718.
65. Yoshida, T., et al. (2006). Systematic doping evolution of the underlying Fermi surface of $La_{2-x}Sr_xCuO_4$. *Physical Review B, 74*, 224510.
66. Xia, Y., et al. (2009). Fermi surface topology and low-lying quasiparticle dynamics of parent $Fe_{1+x}Te/Se$ superconductor. *Physical Review Letters, 103*, 037002.
67. Qian, T., et al. (2011). Absence of a holelike Fermi surface for the iron-based $K_{0.8}Fe_{1.7}Se_2$ superconductor revealed by angle-resolved photoemission spectroscopy. *Physical Review Letters, 106*, 187001.
68. Vaknin, D., et al. (1987). Antiferromagnetism in La_2CuO_{4-y}. *Physical Review Letters, 58*, 2802–2805.
69. Gruner, G. (2000). *Density waves in solids*. Westview Press, Boulder.
70. Fawcett, E. (1988). Spin-density-wave antiferromagnetism in chromium. *Reviews of Modern Physics, 60*, 209–283.
71. Schafgans, A. A., et al. (2012). Electronic correlations and unconventional spectral weight transfer in the high-temperature pnictide $BaFe_{2-x}Co_xAs_2$ superconductor using infrared spectroscopy. *Physical Review Letters, 108*, 147002.
72. Harriger, L. W., et al. (2011). Nematic spin fluid in the tetragonal phase of $BaFe_2As_2$. *Physical Review B, 84*, 054544.
73. Kaneshita, E., & Tohyama, T. (2010). Spin and charge dynamics ruled by antiferromagnetic order in iron pnictide superconductors. *Physical Review B, 82*, 094441.
74. Coldea, R., et al. (2001). Spin waves and electronic interactions in La_2CuO_4. *Physical Review Letters, 86*, 5377–5380.
75. Gretarsson, H., et al. (2011). Revealing the dual nature of magnetism in iron pnictides and iron chalcogenides using X-ray emission spectroscopy. *Physical Review B, 84*, 100509.
76. Borisenko, S. V., et al. (2010). Superconductivity without nesting in LiFeAs. *Physical Review Letters, 105*, 067002.
77. Ishikawa, Y., Noda, Y., Uemura, Y. J., Majkrzak, C. F., & Shirane, G. (1985). Paramagnetic spin fluctuations in the weak itinerant-electron ferromagnet MnSi. *Physical Review B, 31*, 5884–5893.
78. Boeri, L., Dolgov, O. V., & Golubov, A. A. (2008). Is $LaFeAsO_{1-x}F_x$ an electron-phonon superconductor? *Physical Review Letters, 101*, 026403.
79. Moriya, T., & Ueda, K. (2003). Antiferromagnetic spin fluctuation and superconductivity. *Reports on Progress in Physics, 66*, 1299.
80. Scalapino, D. J., Loh, E., & Hirsch, J. E. (1986). d-wave pairing near a spin-density-wave instability. *Physical Review B, 34*, 8190–8192.
81. Tsuei, C. C., & Kirtley, J. R. (2000). Pairing symmetry in cuprate superconductors. *Reviews of Modern Physics, 72*, 969–1016.
82. Bennemann, K.-H., & Ketterson, J. B. (2008). *Superconductivity: volume 1: conventional and unconventional superconductors*. Springer, Berlin.
83. Zhang, C., et al. (2011). Neutron scattering studies of spin excitations in hole-doped $Ba_{0.67}K_{0.33}Fe_2As_2$ superconductor. *Science Reports, 1*, 115.
84. Friemel, G., et al. (2012). Reciprocal-space structure and dispersion of the magnetic resonant mode in the superconducting phase of $Rb_xFe_{2-y}Se_2$ single crystals. *Physical Review B, 85*, 140511.
85. Friemel, G., et al. (2012). Conformity of spin fluctuations in alkali-metal iron selenide superconductors inferred from the observation of a magnetic resonant mode in $K_xFe_{2-y}Se_2$. *Europhysics Letters, 99*, 67004.
86. Taylor, A. E., et al. (2012). Spin-wave excitations and superconducting resonant mode in $Cs_xFe_{2-y}Se_2$. *Physical Review B, 86*, 094528.

87. Onari, S., Kontani, H., & Sato, M. (2010). Structure of neutron-scattering peaks in both s_{++}-wave and $s_{\pm}$-wave states of an iron pnictide superconductor. *Physical Review B*, *81*, 060504.
88. Annett, J. F. (1990). Symmetry of the order parameter for high-temperature superconductivity. *Advances in Physics*, *39*, 83–126.
89. Van Harlingen, D. J. (1995). Phase-sensitive tests of the symmetry of the pairing state in the high-temperature superconductors: evidence for $d_{x^2-y^2}$ symmetry. *Reviews of Modern Physics*, *67*, 515–535.
90. Hanaguri, T., Niitaka, S., Kuroki, K., & Takagi, H. (2010). Unconventional s-wave superconductivity in Fe(Se, Te). *Science*, *328*, 474.
91. Golubov, A. A., & Mazin, I. I. (2013). Designing phase-sensitive tests for Fe-based superconductors. *Applied Physics Letters*, *102*, 032601.
92. Shimojima, T., et al. (2010). Orbital-dependent modifications of electronic structure across the magnetostructural transition in $BaFe_2As_2$. *Physical Review Letters*, *104*, 057002.
93. Sudayama, T., et al. (2011). Doping-dependent and orbital-dependent band renormalization in $Ba(Fe_{1-x}Co_x)_2As_2$ superconductors. *Journal of the Physical Society of Japan*, *80*, 113707.
94. Arham, H. Z., et al. (2012). Detection of orbital fluctuations above the structural transition temperature in the iron pnictides and chalcogenides. *Physical Review B*, *85*, 214515.
95. Kontani, H., & Onari, S. (2010). Orbital-fluctuation-mediated superconductivity in iron pnictides: analysis of the five-orbital Hubbard–Holstein model. *Physical Review Letters*, *104*, 157001.
96. Yamada, T., & Ōno, Y. (2012). Dynamical mean-field study of local pairing interaction mediated by spin and orbital fluctuations in iron pnictide superconductors. arXiv:1209.4954 (unpublished).
97. Hardy, F., et al. (2010). Doping evolution of superconducting gaps and electronic densities of states in $Ba(Fe_{1-x}Co_x)_2As_2$ iron pnictides. *Europhysics Letters*, *91*, 47008.
98. Grimvall, G. (1976). The electron-phonon interaction in normal metals. *Physica Scripta*, *14*, 63.
99. Grimvall, G. (1981). In: E.P. Wohlfarth (Ed.), *The electron–phonon interaction in metals, selected topics in solid state physics*, North-Holland, Amsterdam.
100. Suhl, H., Matthias, B. T., & Walker, L. R. (1959). Bardeen–Cooper–Schrieffer theory of superconductivity in the case of overlapping bands. *Physical Review Letters*, *3*, 552–554.
101. Dolgov, O. V., Mazin, I. I., Parker, D., & Golubov, A. A. (2009). Interband superconductivity: contrasts between Bardeen–Cooper–Schrieffer and Eliashberg theories. *Physical Review B*, *79*, 060502.
102. Nicol, E. J., Carbotte, J. P., & Timusk, T. (1991). Optical conductivity in high-T_c superconductors. *Physical Review B*, *43*, 473–479.
103. Guo, J., et al. (2010). Superconductivity in the iron selenide $K_xFe_2Se_2$ ($0 \leq x \leq 1.0$). *Physical Review B*, *82*, 180520.
104. Ying, J. J., et al. (2011). Superconductivity and magnetic properties of single crystals of $K_{0.75}Fe_{1.66}Se_2$ and $Cs_{0.81}Fe_{1.61}Se_2$. *Physical Review B*, *83*, 212502.
105. Li, C.-H., Shen, B., Han, F., Zhu, X., & Wen, H.-H. (2011). Transport properties and anisotropy of $Rb_{1-x}Fe_{2-y}Se_2$ single crystals. *Physical Review B*, *83*, 184521.
106. Mizuguchi, Y., et al. (2011). Transport properties of the new Fe-based superconductor $K_xFe_2Se_2$ ($T_c = 33$K). *Applied Physics Letters*, *98*, 042511.
107. Wang, A. F., et al. (2011). Superconductivity at 32 K in single-crystalline $Rb_xFe_{2-y}Se_2$. *Physical Review B*, *83*, 060512.
108. Fang, M.-H., et al. (2011). Fe-based superconductivity with $T_c = 31$ K bordering an antiferromagnetic insulator in (Tl, K)Fe_xSe_2. *Europhysics Letters*, *94*, 27009.
109. Zhang, A. M., et al. (2012). Effect of iron content and potassium substitution in $A_{0.8}Fe_{1.6}Se_2$ (A=K, Rb, Tl) superconductors: a Raman scattering investigation. *Physical Review B*, *86*, 134502.
110. Tsurkan, V., et al. (2011). Anisotropic magnetism, superconductivity, and the phase diagram of $Rb_{1-x}Fe_{2-y}Se_2$. *Physical Review B*, *84*, 144520.

111. Ksenofontov, V., et al. (2011). Phase separation in superconducting and antiferromagnetic $Rb_{0.8}Fe_{1.6}Se_2$ probed by Mössbauer spectroscopy. *Physical Review B, 84*, 180508.
112. Shermadini, Z., et al. (2012). Superconducting properties of single-crystalline $A_xFe_{2-y}Se_2$ (A = Rb, K) studied using muon spin spectroscopy. *Physical Review B, 85*, 100501.
113. Texier, Y., et al. (2012). NMR study in the iron-selenide $Rb_{0.74}Fe_{1.6}Se_2$: determination of the superconducting phase as iron vacancy-free $Rb_{0.3}Fe_2Se_2$. *Physical Review Letters, 108*, 237002.
114. Charnukha, A., et al. (2012). Nanoscale layering of antiferromagnetic and superconducting phases in $Rb_2Fe_4Se_5$ single crystals. *Physical Review Letters, 109*, 017003.
115. Liu, Y., Xing, Q., Dennis, K. W., McCallum, R. W., & Lograsso, T. A. (2012). Evolution of precipitate morphology during heat treatment and its implications for the superconductivity in $K_xFe_{1.6+y}Se_2$ single crystals. *Physical Review B, 86*, 144507.
116. Chen, F., et al. (2011). Electronic identification of the parental phases and mesoscopic phase separation of $K_xFe_{2-y}Se_2$ superconductors. *Physical Review X, 1*, 021020.
117. Ricci, A., et al. (2011). Nanoscale phase separation in the iron chalcogenide superconductor $K_{0.8}Fe_{1.6}Se_2$ as seen via scanning nanofocused X-ray diffraction. *Physical Review B, 84*, 060511.
118. Cai, P., et al. (2012). Imaging the coexistence of a superconducting phase and a charge-density modulation in the $K_{0.73}Fe_{1.67}Se_2$ superconductor using a scanning tunneling microscope. *Physical Review B, 85*, 094512.
119. Charnukha, A., et al. (2012). Optical conductivity of superconducting $Rb_2Fe_4Se_5$ single crystals. *Physical Review B, 85*, 100504.
120. Homes, C. C., Xu, Z. J., Wen, J. S., & Gu, G. D. (2012). Optical conductivity of superconducting $K_{0.8}Fe_{2-y}Se_2$ single crystals: evidence for a Josephson-coupled phase. *Physical Review B, 85*, 180510.
121. Homes, C. C., Xu, Z. J., Wen, J. S., & Gu, G. D. (2012). Effective medium approximation and the complex optical properties of the inhomogeneous superconductor $K_{0.8}Fe_{2-y}Se_2$. *Physical Review B, 86*, 144530.
122. Wang, C. N., et al. (2012). Macroscopic phase segregation in superconducting $K_{0.73}Fe_{1.67}Se_2$ as seen by muon spin rotation and infrared spectroscopy. *Physical Review B, 85*, 214503.
123. Lazarević, N., et al. (2012). Vacancy-induced nanoscale phase separation in $K_xFe_{2-y}Se_2$ single crystals evidenced by Raman scattering and powder X-ray diffraction. *Physical Review B, 86*, 054503.
124. Li, W., et al. (2012). Phase separation and magnetic order in K-doped iron selenide superconductor. *Nature Physics, 8*, 126–130.
125. Li, W., et al. (2012). KFe_2Se_2 is the parent compound of K-doped iron selenide superconductors. *Physical Review Letters, 109*, 057003.
126. Simonelli, L., et al. (2012). Coexistence of different electronic phases in the $K_{0.8}Fe_{1.6}Se_2$ superconductor: a bulk-sensitive hard X-ray spectroscopy study. *Physical Review Letters, 85*, 224510.
127. Weyeneth, S., et al. (2012). Superconductivity and magnetism in $Rb_xFe_{2-y}Se_2$: impact of thermal treatment on mesoscopic phase separation. *Physical Review B, 86*, 134530.
128. Yuan, R. H., et al. (2012). Nanoscale phase separation of antiferromagnetic order and superconductivity in $K_{0.75}Fe_{1.75}Se_2$. *Science Reports, 2*, 221.
129. Shoemaker, D. P., et al. (2012). Phase relations in $K_xFe_{2-y}Se_2$ and the structure of superconducting $K_xFe_2Se_2$ via high-resolution synchrotron diffraction. *Physical Review B, 86*, 184511.
130. Wen, H.-H. (2012). Overview on the physics and materials of the new superconductor $K_xFe_{2-y}Se_2$. *Reports on Progress in Physics, 75*, 112501.
131. Yan, X.-W., Gao, M., Lu, Z.-Y., & Xiang, T. (2011). Ternary iron selenide $K_{0.8}Fe_{1.6}Se_2$ is an antiferromagnetic semiconductor. *Physical Review B, 83*, 233205.
132. Xu, M., et al. (2012). Evidence for an s-wave superconducting gap in $K_xFe_{2-y}Se_2$ from angle-resolved photoemission. *Physical Review B, 85*, 220504.

133. Bao, W., et al. (2011). A novel large moment antiferromagnetic order in $K_{0.8}Fe_{1.6}Se_2$ superconductor. *Chinese Physics Letters, 28*, 086104.
134. Wang, Z., et al. (2011). Microstructure and ordering of iron vacancies in the superconductor system $K_yFe_xSe_2$ as seen via transmission electron microscopy. *Physical Review B, 83*, 140505.
135. Maier, T. A., Graser, S., Hirschfeld, P. J., & Scalapino, D. J. (2011). d-wave pairing from spin fluctuations in the $K_xFe_{2-y}Se_2$ superconductors. *Physical Review B, 83*, 100515.
136. Saito, T., Onari, S., & Kontani, H. (2011). Emergence of fully gapped s_{++}-wave and nodal d-wave states mediated by orbital and spin fluctuations in a ten-orbital model of KFe_2Se_2. *Physical Review B, 83*, 140512.
137. Wang, F., et al. (2011). The electron pairing of $K_xFe_{2-y}Se_2$. *Europhysics Letters, 93*, 57003.
138. May, A. F., et al. (2012). Spin reorientation in $TlFe_{1.6}Se_2$ with complete vacancy ordering. *Physical Review Letters, 109*, 077003.
139. Yan, X.-W., & Gao, M. (2012). The effect of the Wyckoff position of the K atom on the crystal structure and electronic properties of the compound KFe_2Se_2. *Journal of Physics: Condensed Matter, 24*, 455702.

Chapter 3
Experimental and Theoretical Methods

> *In theory, there is no difference between theory and practice. In practice, there is.*
>
> —Jan L. A. van de Snepscheut

3.1 Spectroscopic Ellipsometry

Optical spectroscopy is one of the oldest techniques in condensed-matter research, starting from the earliest observations of interference fringes in the intensity of light scattered from thin films, dating back to Newton, to taking into account the relative phases of two independent polarizations of light, carried out in the work of Drude. The former observations only involved the analysis of light intensity and laid the foundation of modern reflectance measurements, one of the most popular optical spectroscopic techniques in the world. In its essence the technique involves a simple measurement of the intensity of light reflected from a sample surface $I(\omega)$ at angles of incidence very close to the normal to the sample surface and at various frequencies ω of the incident light. Since electromagnetic radiation is inherently transverse (i.e. the polarization of both the electric and the magnetic field is perpendicular to the propagation vector), in the reflectance geometry both fields in the sample oscillate almost parallel to the sample surface. This feature of the reflectance geometry proves very useful when studying strongly anisotropic samples because to a high degree of accuracy only the in-plane response of the material is probed, without contamination from the out-of-plane component. The optical response of a nonmagnetic linear material is fully characterized by the so-called dielectric tensor $\varepsilon_{ij}(\omega),\ i, j = x, y, z$, where ω is the frequency of the probing light and all ε are complex-valued functions of frequency, which can be diagonalized to reduce to just three complex functions of frequency $(\varepsilon_x(\omega), \varepsilon_y(\omega), \varepsilon_z(\omega))$, where $\varepsilon = \varepsilon_1 + i\varepsilon_2$ and $\varepsilon_{1,2}$ are both real. In the reflectance configuration predominantly the in-plane, or x, y, components of the dielectric tensor

A. Charnukha, *Charge Dynamics in 122 Iron-Based Superconductors*,
Springer Theses, DOI: 10.1007/978-3-319-01192-9_3,

contribute to the overall optical response.[1] For sufficiently large samples, reflectance measurements can be extended to very low energies, down to 15 cm^{-1} or about 2 meV, particularly using intense synchrotron radiation as a light source.

The reflectance technique does, however, have an important limitation: as only the intensity of reflected light is detected (one real-valued function of frequency), even in the case of a sample with a cubic crystallographic symmetry ($\varepsilon_x = \varepsilon_y = \varepsilon_z = \varepsilon$) the single *complex* function of frequency with two independent real components cannot be extracted based on the measured data alone. Fortunately, the dielectric function is not an arbitrary function but is a response function,[2] that is, it describes the response of a system to a certain perturbation. In other words, $\varepsilon(\omega) - 1$ is bounded by the condition of causality, i.e. the future state of a material cannot influence its present state, a condition, which turns out to be sufficient to guarantee the analyticity of $\varepsilon(\omega) - 1$ in the upper complex half-plane $\delta > 0$ of the complex frequency $\omega + i\delta$. It follows then [1] that the real and the imaginary part of the complex dielectric function are not independent but rather connected by the so-called KK relations:

$$\begin{aligned} \varepsilon_1(\omega) - 1 &= \frac{1}{\pi}\mathscr{P}\int_{-\infty}^{\infty} \frac{\varepsilon_2(\Omega)}{\Omega - \omega} d\Omega, \\ \varepsilon_2(\omega) &= -\frac{1}{\pi}\mathscr{P}\int_{-\infty}^{\infty} \frac{\varepsilon_1(\Omega) - 1}{\Omega - \omega} d\Omega, \end{aligned} \tag{3.1}$$

where $\mathscr{P}$ denotes the Cauchy principal value of the integral to handle the singularity with respect to frequency. These relations imply that all the optical properties contained in the complex dielectric function are actually captured by either of its components alone, and one can be obtained from the other using Eq. (3.1). In the above form the KK relations are not suitable for application to physical measurements because they assume that the response function is known at negative frequencies. However, since any physical response function $\chi(\omega)$ is the Fourier transform of $\chi(t - t')$, it follows that $\chi(-\omega) = \chi^*(\omega)$. Therefore, $\chi_1(\omega)$ is an even and $\chi_2(\omega)$—an odd function of frequency. One can use this fact to reduce Eq. (3.1) to integrals of definite parity. To that end we multiply the numerator and denominator of both equations by $\Omega + \omega$ and use the parity of $\varepsilon_{1,2}(\omega)$ to arrive at:

$$\begin{aligned} \varepsilon_1(\omega) - 1 &= \frac{2}{\pi}\mathscr{P}\int_{0}^{\infty} \frac{\Omega\varepsilon_2(\Omega)}{\Omega^2 - \omega^2} d\Omega, \\ \varepsilon_2(\omega) &= -\frac{2\omega}{\pi}\mathscr{P}\int_{0}^{\infty} \frac{\varepsilon_1(\Omega) - 1}{\Omega^2 - \omega^2} d\Omega. \end{aligned} \tag{3.2}$$

[1] A small contamination from the z component of the dielectric tensor is always present due to an inevitable deviation of the angle of incidence from the normal to the sample surface arising from the simple practical limitation that the incident and reflected light must be spatially separated to enable the analysis of the latter.

[2] This is, strictly speaking, only true when the unity contribution to the real part of the dielectric function has been subtracted, that is $\varepsilon(\omega) - 1$ is a response function. See below for the definition of the dielectric function in terms of the charge susceptibility, which is a true response function.

The KK relations (3.2) imply that measurement of the intensity of reflected light is sufficient to extract the complete complex dielectric function of a material. This conclusion holds true in theory but in practice its application is limited by the fact that one can never measure the intensity of reflected light in the entire spectral range $\omega \in [0, \infty)$. As it was mentioned above, a synchrotron-based reflectance setup can achieve frequencies as low as 2 meV and many achieve frequencies as high as 7 eV. To apply the KK relations (3.2) one must resort to extrapolation of $I(\omega)$ to 0 and ∞ using more or less justified physical models. The uncertainty of the extrapolation at low and high frequencies leads to errors in the extracted $\varepsilon_{1,2}(\omega)$ particularly at frequencies close to the ends of the experimental spectral range. Far away from the extrapolation regions the results of such analysis only weakly depend on the type of the extrapolation function used (see Sect. 3.5). The simplicity of the practical realization and the minimal contamination of the in-plane response by the out-of-plane optical properties, have made this technique the first choice for optical studies. It has been perfected over its long history and is now used in many condensed-matter laboratories around the world.

The technique of spectroscopic ellipsometry, which is the focus of this section, traces its history back to the early work of Drude, who used the phase shift between two mutually perpendicular components of light polarization, introduced upon reflection from a sample, to measure thicknesses down to 1 Å. Since the polarization of light with the two components out of phase is in general elliptical, the technique derived from these seminal measurements came to be called 'ellipsometry'. The technique rose to fame after the invention of the transistor and the development of integrated circuits, which required high-quality and highly homogeneous silicon wafers. Ellipsometry was found to be the simplest and most reliable tool for the thickness control of industrially produced silicon wafers to very high accuracy. This industrially important application generated tremendous interest to the technique and led to the fast development of the field and industrialization of the obtained knowledge. Today a number of companies provide state-of-the-art ellipsometers for industry (Woollam, SENTECH, Angstrom Advanced to name a few). The current understanding of the technique and its applications is brought together in probably the most detailed and up-to-date scholarly treatise 'Handbook of ellipsometry' [2].

A typical ellipsometry setup is composed of the following basic components (illustrated schematically in Fig. 3.1): light source, polarizer, compensator (optional), sample, compensator (optional), analyzer, detector. Let us discuss the function and realization of all these components. The light source generally generates broadband unpolarized light and can be thermal (various incandescent electric lamps) or use the synchrotron radiation of accelerated electrons. Since these sources provide broadband light, additional processing is required to achieve wavelength-specific analysis. To this end, there are two types of experimental configurations: single-wavelength and Fourier-transform spectroscopy. In the former a monochromator is used to select a single, well-defined frequency of light, which is then fed into the ellipsometer as shown in Fig. 3.1. In this case the spectral resolution of the system is determined by the quality of the monochromator, i.e. how narrow the frequency distribution is around the one selected nominally.

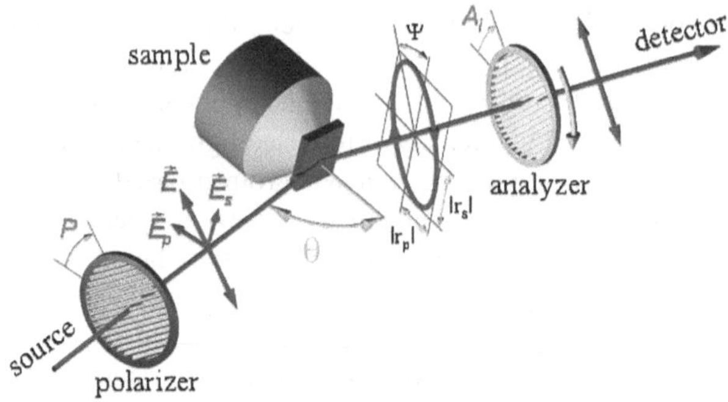

Fig. 3.1 Simplest ellipsometry configuration: light from the source is linearly polarized at an angle P with respect to the plane of incidence and is reflected from the sample at an angle of incidence θ, which in general produces elliptically polarized light according to Eq. (3.3); this elliptical polarization is then recorded at various analyzer angles A_i by the detector. Image courtesy A. V. Boris, Max Planck Institute for Solid-State Research, 70569 Stuttgart, Germany, Heisenbergstr. 1

In the Fourier-transform configuration, on the other hand, all frequencies of the broadband light source are investigated simultaneously by using a Michelson interferometer to modulate the broadband light. This way spectral information is converted into the temporal profile of the outgoing light and is later recovered at the detector by performing the Fourier transform of the recorded time-dependent intensity. In this case the spectral resolution is determined by the amount of detail contained in the temporal profile of light at large times, which translates into the length of the path traversed by the moving mirror of the Michelson interferometer, because the difference between two very similar frequencies can only be recognized at very large times/distances. The two configurations of the light source show a very similar level of performance in state-of-the-art ellipsometric systems. One of the advantages of the single-wavelength approach, however, is the ability to perform temperature scans at a given frequency of light without recording and processing all other frequencies (redundant in this case) as would be the case in Fourier-transform spectroscopy.

Further operation of the ellipsometer is independent of the type of the light source used. The output of the source is first linearly polarized to provide a well-defined initial polarization state for the system. In some cases, as will be discussed below, no linear polarization can provide sufficient sensitivity of the ellipsometer to the optical properties of the sample and a more general, elliptical, polarization of incident light must be used. In the simplest case of circular polarization this transformation of the polarization state is achieved by placing a quarter-wave plate ($\lambda/4$) in the beam path. This and any other optical component modifying the polarization of light by retarding one of the polarization components is called a retarder (compensator). The light with a well-defined polarization state is then reflected from a sample, which changes its polarization state according to the classical Fresnel equations (see below). This new polarization state is then analyzed by probing the intensity of its various polarization components using another polarizer as an analyzer and a detector. An additional compensator may be installed between the sample and the analyzer for more detailed measurements (see Ref. [2]).

The physical realization of the aforementioned components largely depends on the spectral range, in which investigations are to be carried out. Spectral limitations of light sources, polarizers, compensators, and detectors dictate the appropriate choice of the optical components. For instance, the frequencies in the upper visible and lower ultraviolet range of the spectrum are usually detected by a photomultiplier tube, those in the lower visible and middle-infrared spectral range—by a semiconductor detector, and to detect even lower frequencies one has to resort to bolometers. Ones of the most important conditions to be met when choosing polarizers is the degree of the rejection of unwanted polarization and the accuracy and reproducibility of the rotation stages, on which the polarizers are mounted, as the precise knowledge of the polarization state is of utmost importance for ellipsometric measurements.

Another concern to heed is the quality of the vacuum in a cryostat. As the temperature of the sample is lowered it will eventually reach the sublimation temperature of water ice, which means that below that temperature a film of ice will have condensed on the sample surface. This sublimation temperature is rather high even at the lowest realistically achievable pressures: a pressure of about 10^{-11} mbar results in a sublimation temperature of roughly 125 K [3, 4], much higher than the base temperature of liquid-helium–flow cryostats, typically 5–6 K. This means that the formation of an ice layer is inevitable and its thickness is determined by the cryostat pressure. This limitation becomes particularly severe for measurements in the visible and even more so in the ultraviolet part of the spectral range because as the wavelength of light decreases, thinner and thinner layers of ice will significantly distort experimental results. Therefore, for measurements in this spectral range low-temperature pressures on the order of 10^{-10} mbar are required. To illustrate the effect of vacuum conditions on the formation of ice on the sample, we estimate the upper bound on the formation rate of molecular ice layers. We assume that the sample is at a temperature below the sublimation temperature of water ice at the respective pressure and that the cryostat is at room temperature. The pressure of a molecular gas can be obtained as $P = F/S = \Delta p/\Delta t = m\bar{v}\Delta N/\Delta t = m\bar{v}n/S$, where P is the cryostat pressure, defined as a force F acting on a surface S, Δp is the momentum change due to a collision with a water molecule, Δt is the time interval considered, ΔN is the number of collisions within that interval, $n = \Delta N/\Delta t$ is the collision rate (water molecules per second), and $\bar{v} = \sqrt{8k_{\mathrm{B}}T/(\pi m)}$ is the Maxwellian average velocity, in which k_{B} is the Boltzmann constant, T is the cryostat gas temperature, and m is the mass of a water molecule. Here we assume that all water molecules colliding with a sample get stuck to its surface and thus transfer all of their momentum to it ($\Delta p = m\bar{v}$), leading to the following layer formation rate: $n_{\mathrm{layer}} = PS_{\mathrm{u.c.}}\sqrt{\pi}/\sqrt{8mk_{\mathrm{B}}T}$. Approximating the unit-cell area of water ice per water molecule to $S_{\mathrm{u.c.}} = 25\ \text{Å}^2$ and assuming a cryostat water pressure of 10^{-11} mbar $= 10^{-9}$ Pa one arrives at the formation time of one layer ($1/n_{\mathrm{layer}}$) of about 15 h. This is the upper bound for the water ice formation rate and in reality the rate can be significantly smaller. If the cryostat water pressure deteriorates to 10^{-8} mbar the formation time of one layer of ice decreases to 1 min.

This implies that by the end of typical measurement time of an hour 60 layers of ice with the total thickness of about 50 nm will have formed. Such a thickness of ice is certainly sufficient to significantly affect the results of an ellipsometric measurement, particularly in the visible and ultraviolet spectral range.

3.1.1 Formalism

Now that the components of the ellipsometer and their realization have been discussed, we turn to the formalism and data analysis techniques. First of all, as mentioned above, the initial well-defined polarization state of light is changed upon reflection from a sample. This change is determined by the optical properties of the reflecting medium and in the simplest case of an isotropic ($\varepsilon_j = \varepsilon = \varepsilon_1 + i\varepsilon_2,\ j = $ x,y,z) semi-infinite homogeneous non-depolarizing medium in the reflection configuration shown in Fig. 3.2 is given by the celebrated Fresnel equations [5]:

$$\frac{E_{\mathrm{s,r}}}{E_{\mathrm{s,i}}} = \frac{\cos\theta - \sqrt{\varepsilon - \sin^2\theta}}{\cos\theta + \sqrt{\varepsilon - \sin^2\theta}}, \tag{3.3}$$

$$\frac{E_{\mathrm{p,r}}}{E_{\mathrm{p,i}}} = -\frac{\varepsilon\cos\theta - \sqrt{\varepsilon - \sin^2\theta}}{\varepsilon\cos\theta + \sqrt{\varepsilon - \sin^2\theta}},$$

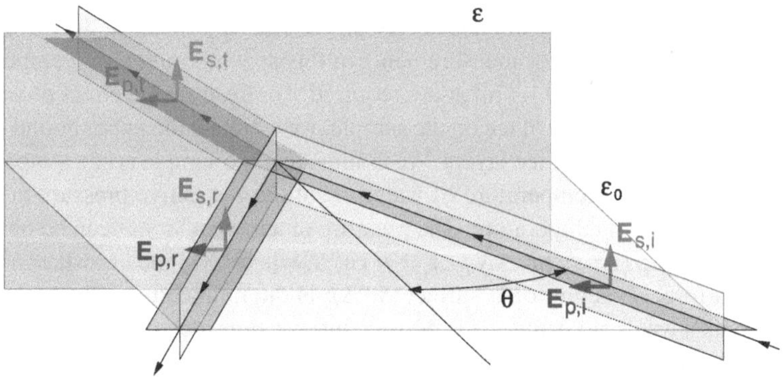

Fig. 3.2 General configuration for light incidence, reflection, and transmission described by Eqs. (3.3). The light is incident (index i) onto the interface between two media: the direction of propagation is from a medium with the dielectric function $\varepsilon_0(\omega)$ (for vacuum $\varepsilon_0(\omega) = 1$ and this case is assumed here) to a medium with the dielectric function $\varepsilon(\omega)$. At the interface, the light is reflected from (index r) and transmitted through (index t) the interface. The so-called plane of incidence is spanned by the propagation vector of the incoming light and the normal to the interface (the propagation vector of the reflected and transmitted light lies in the same plane). Any polarization of light can be decomposed into two independent and mutually orthogonal components: one in the plane of incidence (index p, *red color*), the other perpendicular to it (index s, *blue color*). For non-depolarizing media these two components of light polarization transform independently upon reflection/transmission and do not mix, as illustrated by Eqs. (3.3). The angle of incidence is denoted by θ

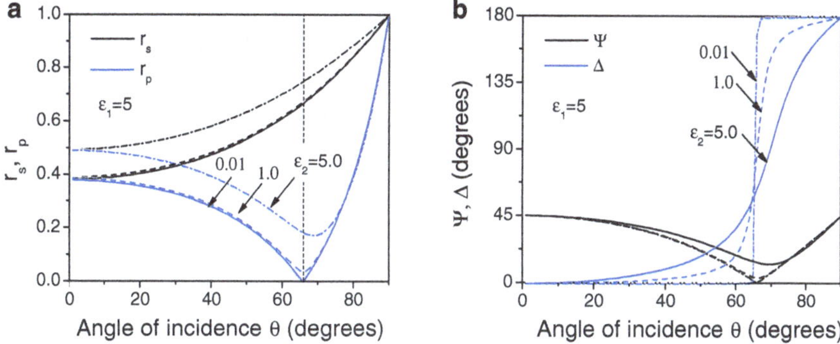

Fig. 3.3 **a** Fresnel amplitudes $|r_{s,p}|$ and **b** ellipsometric angles Ψ, Δ as functions of the angle of incidence θ for $\varepsilon_1 = 5$ and $\varepsilon_2 = 5, 1, 0.01$. The *dashed vertical line* in **a** marks the Brewster angle. For a very detailed discussion of the ellipsometric angles and their dependence on various experimental parameters see Ref. [2]

where $E_{s,r}$ ($E_{s,i}$) is the amplitude of the reflected (incident) electric field polarized perpendicular ('s' comes from the German word 'senkrecht', which means 'perpendicular') to the plane of incidence (see Fig. 3.2), $E_{p,r}$ ($E_{p,i}$) is that polarized in the plane of incidence, and θ is the angle of incidence defined as the angle between the propagation direction of the incident light and the normal to the reflecting surface, as shown in Fig. 3.2. Since ε is a complex quantity, so are the ratios $r_a = E_{a,r}/E_{a,i}$, $a =$ s,p. Rather than using their real and imaginary parts, one can equivalently represent them by their real-valued amplitudes and phases: $r_a = |r_a| \exp(i\delta_a)$, $a = s, p$. Thus by measuring the ratios r_s, r_p one can extract the complex dielectric function ε of the reflecting medium. It will be shown in what follows that, in fact, only the ratio of these quantities $\rho = r_p/r_s = \tan\Psi \exp(i\Delta)$ can be obtained. Here we have introduced the conventional ellipsometric notation $\tan(\Psi) = |r_p/r_s|$ and $\Delta = \delta_p - \delta_s$. In the case above this presents no additional problem since a single complex quantity ε can be uniquely determined from ρ. In more complex cases, for instance, in the case of an anisotropic medium with $\varepsilon_x \neq \varepsilon_y \neq \varepsilon_z$, where more than one complex quantity determines the optical response of a medium, several ellipsometric measurements at different angles of incidence and/or in different geometrical configurations must be carried out to provide a sufficient number of experimentally obtained quantities to extract the complete dielectric function. In cases where already Eq. (3.3) provide an adequate description, one has to invert these equations to extract ε from the measured ρ. To make the respective calculations computationally less expensive, an analytic inversion of the Fresnel equations can be used [6].

One of the key issues in spectroscopic ellipsometry is the selection of an adequate angle of incidence at which the measurements should be carried out. This selection turns out to be much more important than the exact (very small) angle of incidence used in the reflectivity configuration. To fully appreciate it let us consider the Fresnel amplitudes $r_{s,p}$ and ellipsometric angles Ψ, Δ as functions of the angle of incidence using Eq. (3.3) for several values of the complex dielectric constant, as shown in Fig. 3.3. Both Fresnel amplitudes start from the same value at small and end up having the same value at large angles of incidence. However, while r_s is a monotonic function of the angle of incidence (black lines in Fig. 3.3a), r_p (blue lines in Fig. 3.3a) has a pronounced minimum at a certain angle, called the Brewster angle. In the absence of absorption ($\varepsilon_2 \approx 0$) r_p goes to zero at this angle, which leads to fully polarized reflected light even when the incident light was unpolarized. If we now consider the corresponding angular dependence of Ψ and Δ, presented in Fig. 3.3b, it becomes evident that ellipsometry is functional only at angles of incidence around the nominal Brewster angle of a material (for some materials absorption can be so strong that the minimum in r_p becomes quite weak). This conclusion follows immediately from the observation that at angles of incidence much smaller or much larger than the Brewster angle both Ψ and Δ become very close to their extremal values (45° for Ψ and 0 or 180° for Δ) and practically do not depend on the values of the dielectric function, making the extraction of the latter from experimentally obtained Ψ and Δ impossible. The limitation of the angles of incidence to rather large values of the Brewster angle, which can be close to 90° for very metallic samples (typical values lie around 80°), significantly reduces the effective sample area probed by the beam to $\approx S\cos\theta$, where S is the total sample area and θ is the angle of incidence. In the reflectivity configuration ($\theta \approx 15°$) the effective area is 97 % of S, while in the ellipsometric configuration ($\theta \approx 80°$) it is only 17 % of S, resulting in a six times lower intensity. At this seemingly prohibitive cost ellipsometry offers several very significant advantages over reflectivity: it provides both the real and the imaginary parts of the dielectric function in absolute units without resorting to the KK relations (3.2) and does not require any reference measurements, thus making the experiment more robust and eliminating the need for irreversible evaporative coating of the sample with gold.

Let us now consider how the ellipsometric angles Ψ and Δ can be obtained by measuring only the intensity of light for different analyzer angles in the configuration shown in Fig. 3.1. We assume that the light incident on the sample is fully linearly polarized by the first polarizer and all the angles are as shown in Fig. 3.1. The polarization of the incident light can be decomposed into two independent components parallel and perpendicular to the plane of incidence, as discussed above: $E_{\mathrm{p,i}} = E_{\mathrm{pol}}\cos P,\ E_{\mathrm{s,i}} = E_{\mathrm{pol}}\sin P$. The corresponding field amplitudes of the light reflected from the sample can be obtained by multiplying the amplitudes of the incident light by the complex reflectivities determined by the sample properties:

$$E_{\mathrm{p,r}} = r_{\mathrm{p}}E_{\mathrm{p,i}} = r_{\mathrm{p}}E_{\mathrm{pol}}\cos P,$$
$$E_{\mathrm{s,r}} = r_{\mathrm{s}}E_{\mathrm{s,i}} = r_{\mathrm{s}}E_{\mathrm{pol}}\sin P.$$

The subsequent passing through the analyzer combines these two components in the direction of the analyzer transmission axis, which results in the following field amplitude at the detector:

$$E_{\rm det} = E_{\rm p,r}\cos A + E_{\rm s,r}\sin P = r_{\rm p}E_{\rm pol}\cos P\cos A + r_{\rm s}E_{\rm pol}\sin P\sin A$$
$$= E_{\rm pol}r_{\rm s}(\rho\cos P\cos A + \sin P\sin A),$$

where, as before, $\rho = \tan\Psi\exp(i\Delta) = r_{\rm p}/r_{\rm s}$. The detector is, of course, only sensitive to the intensity of light I, which is proportional to $|E|^2$:

$$I = I_0\,|r_{\rm s}|^2\cos^2 P\left(\tan^2\Psi\cos^2 A + \tan^2 P\cos^2 A + 2\tan\Psi\cos\Delta\cos A\sin A\right)$$
$$= I_0\,|r_{\rm s}|^2\cos^2 P\left(\tan^2\Psi + \tan^2 P\right)(1 + \alpha\cos 2A + \beta\sin 2A)\,,$$

where

$$\alpha = \frac{\tan^2\Psi - \tan^2 P}{\tan^2\Psi + \tan^2 P},\quad \beta = \frac{2\tan P\tan\Psi\cos\Delta}{\tan^2\Psi + \tan^2 P}. \tag{3.4}$$

The real numbers α and β are the second-order sine and cosine Fourier coefficients of $I(A)$ normalized to the zeroth-order term. Relations (3.4) can be inverted with respect to Ψ and Δ:

$$\tan\Psi = |\tan P|\sqrt{\frac{1+\alpha}{1-\alpha}},\quad \cos\Delta = \mathrm{sgn}(P)\frac{\beta}{\sqrt{1-\alpha^2}}, \tag{3.5}$$

with $\mathrm{sgn}(P)$ denoting the sign of the polarizer angle. Thus to obtain Ψ and Δ one needs to carry out multiple measurements of light intensity in the configuration depicted in Fig. 3.1, take the Fourier transform of this intensity as a function of the analyzer angle, normalize the second-order sine and cosine coefficients by the constant zeroth-order term, and finally use Eq. (3.5) to extract the ellipsometric angles. To obtain the first three orders of the Fourier transform at least three measurements at different analyzer angles are required. Since intensities detected at different analyzer angles are largely uncorrelated and thus are statistically independent random quantities with their statistical properties connected via a deterministic periodic function to those of a set of just a few random quantities common to them (most importantly fluctuating source intensity but also fluctuating sample properties due to its fluctuating temperature etc.), taking the Fourier transform of measurements obtained at progressively more analyzer angles narrows down the statistical distribution of the Fourier coefficients according to the law of big numbers: $\sigma_N = \sigma_0/\sqrt{N}$. In other words, increasing the number of analyzer angles by a certain factor has the same statistical effect as increasing the number of spectra averaged at each analyzer angle by the same factor. In practice the latter approach is usually adopted, with the number of analyzer angles limited to between four and twenty.

3.2 Scattering-Type Scanning Near-Field Optical Microscopy

Spectroscopic ellipsometry and reflectivity can only determine the optical conductivity of a material averaged over the surface area covered by the light beam. It means that in the case of complex materials with interwoven domains of different phases one can only extract an effective response of the system. Determination of the optical properties of each constituent phase requires the precise knowledge of the size, shape, and distribution of the domains, as well as at least some information about the optical properties of the phases. Such information can only be obtained with a spatially resolved microscopy technique, such as STM/STS, which is capable of determining the density of states (DOS) with atomic resolution, nanofocused X-ray diffraction, which provides structural information with a nanoscale spatial resolution etc. If apart from the geometrical, spatial structure of the phase domains their dielectric properties need to be determined then a combination of a microscope and a spectrometer is required. Such an experimental technique has recently been developed [7] (for a scholarly text see Ref. [8]) and bears the name of apertureless or scattering-type scanning near-field spectroscopy (s-SNOM). It combines AFM with an asymmetric Michelson interferometer so that one of the arms of the interferometer is formed by the AFM tip, as shown in Fig. 3.4a. The setup makes use of a powerful laser source operating at a single frequency in the microscopy regime or at several equidistant frequencies (frequency comb) in the spectroscopy regime. Recently, the feasibility of s-SNOM with a thermal broadband source has been demonstrated [9]. The light from the source is fed into an asymmetric Michelson interferometer with

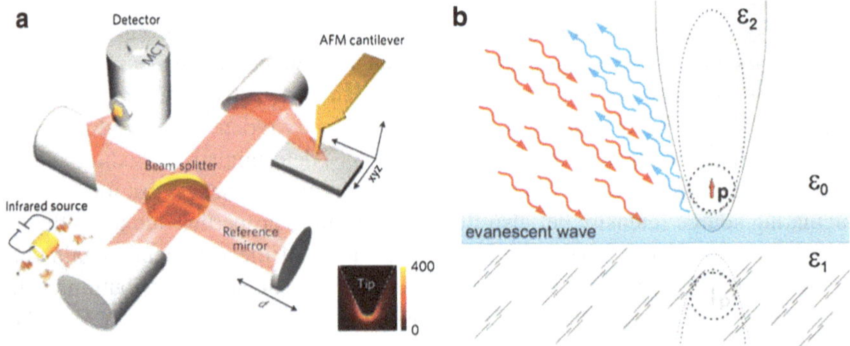

Fig. 3.4 **a** Schematic representation of a typical apertureless (scattering-type) s-SNOM microscope/spectrometer. The *inset* shows a simulation of near-field intensity in the vicinity of an AFM tip. Reprinted by permission from Macmillan Publishers Ltd: Nature Materials [9], copyright (2011). **b** Near-field distribution of the electromagnetic field between the sample surface and the probing tip is strongly affected by the optical properties of both materials, ε_0 and ε_1, as well as those of the tip, ε_2. The evanescent, exponentially decaying, near-field wave in the vicinity of the sample surface is picked up by the tip and scattered into the far-field region, similar to the operation of an antenna [7, 8]. Near-field intensity between the sample surface and the tip can be modeled as resulting from the interaction of an extended (*dotted oval*) or a point (*dashed circle*) dipole in the tip and its electrostatic image in the sample [10]

one arm formed by a moving mirror and the other one by the probing AFM tip. Thus the transferred to the far-field region by the antenna-action of the tip light returns to the interferometer, in which it is recombined with the source light at the beam splitter. The resulting interference pattern is recorded by the detector. The spatial resolution of an s-SNOM setup is determined by the radius of the AFM tip used, with industrially produced and routinely available tips featuring radii as small as 10 nm. The detection scheme usually operates in the so-called pseudoheterodyne interferometric mode, in which the tip is driven at a fixed oscillation frequency and the resulting periodic signal recorded by the detector is demodulated at a chosen harmonic to reduce contaminating background signal and noise.

Employing an AFM microscope in one of the arms of an interferometer allows one to record the topography profile of the sample surface with a typical AFM sensitivity of less than 1 nm and lateral spatial resolution below 10 nm as well as obtain the optical amplitude and phase at each point of the surface map simultaneously. These amplitude and phase, complemented by the reference measurement of a material with known optical properties, can then be analyzed in an appropriate model of the near-field sample-tip interaction to extract the optical conductivity of the sample. The simplest interaction model is the point dipole model shown in Fig. 3.4b (dashed circles), in which both the tip and the reaction of the surface are represented as point dipoles. This model captures the main qualitative features of the near-field signal but fails on a quantitative level. To remedy this situation, a more elaborate but still analytically solvable model has been put forward [10], which treats the tip as an extended dipole (Fig. 3.4b, dotted oval) with two separate charges and the corresponding image in the sample to evaluate the near-field interaction. This model has demonstrated quantitative agreement with experiment and was used for the analysis of the data in this thesis, as will be discussed in Sect. 4.4.

3.3 Low-Energy Muon-Spin Rotation/Relaxation

Given the proximity of superconductivity to antiferromagnetism and the very likely possibility that magnetic interactions mediate superconducting pairing in the iron-based superconductors, it becomes indispensable to achieve a detailed understanding of their magnetic properties in order to gain further insight into the mechanism of superconductivity in these materials. There are many powerful and largely complementary techniques that allow one to investigate magnetic properties and the effect that superconductivity exerts on them. The most direct probe is arguably neutron scattering, which is capable of mapping out the magnetic ground state as well as its elementary excitations in reciprocal space. Magnetism in real space can be studied by NMR, Mössbauer spectroscopy, and μSR to name a few. In this work we have used the low-energy variation of the standard μSR technique to study the strongly magnetic phase-separated iron-selenide superconductors, as will be discussed in detail in Sect. 4.4. LE-μSR features several important advantages over other approaches:

- unlike in neutron scattering, the magnetic moment and the magnetic volume fraction of a material can be determined separately,
- unlike in NMR, no magnetic field or low temperatures are required for operation and thus remain free control parameters,
- in the case of LE-μSR, the implantation depth of muons in a sample can be controlled, enabling the investigation of magnetic and superconducting properties as a function of distance from the sample surface. This capability is particularly useful when studying thin films because by choosing an appropriate shallow muon stopping profile the contribution of the substrate can be eliminated.

Since the leading magnetic interaction takes place between magnetic dipole moments generated by spins, it is important to ascertain full polarization of the probe's spin for a reliable extraction of the magnetic properties of the medium under investigation. In the case of NMR, a large degree of polarization of the atomic nuclei is achieved by means of very low sample temperatures (to minimize the thermal excitation of nuclear spins) and large magnetic fields. For the μSR technique both requirements become irrelevant due to a unique property of the muon production process, from which muons emerge with a very large degree of polarization (almost 100 %). Although muons in principle can be produced in a variety of processes, *low-energy* muons that would stop in a sample of a reasonable size can only be obtained in a pion decay [11]. Pions themselves can be routinely produced in collisions of high-energy protons (>500 MeV) with the atoms of a target made up of a light element, such as carbon or beryllium. The pions produced in this reaction quickly decay into a muon and a muon (anti)neutrino via $\pi^- \rightarrow \mu^- + \bar{\nu}_\mu$ and $\pi^+ \rightarrow \mu^+ + \nu_\mu$. Due to the maximal parity violation in the weak-interaction–mediated pion decay, the emerging muon is almost completely spin-polarized in the direction opposite its momentum. In practice, almost all muon sources produce positive muons because they can be obtained from pions at rest in the surface layer of the target (surface muons) and thus have a lower energy and are stable against nuclear capture in the sample. A positive surface muon produced in the aforementioned pion decay has an energy of 4 MeV and a lifetime of 2.2 μs, after which it emits a positron in a weak-interaction-mediated and thus parity-violating decay preferentially along the direction of its spin. The statistical polarization of an ensemble of muons can be traced in real time by collecting and analyzing decay positrons. Since the original muon beam is fully spin-polarized all variations in its spin direction with time must result from the interaction with the magnetic moments of ions in the material under study. The feasibility of μSR, therefore, is based on parity violation in the weak-interaction-mediated decay of the pion and the muon.

In the conventional μSR technique energetic 4 MeV muons are used directly to probe the magnetic properties of a sample: depending on the sample density muons would stop at an interstitial site within 0.1–1 mm from the sample surface. This allows one to study bulk magnetic properties. In the LE-μSR technique, on the other hand, the original beam of muons is first slowed down to thermal velocities in a moderating layer of a solid noble gas, as shown in Fig. 3.5, and then accelerated by a controlled electric field to desired energies. The choice of the final energy of a muon upon

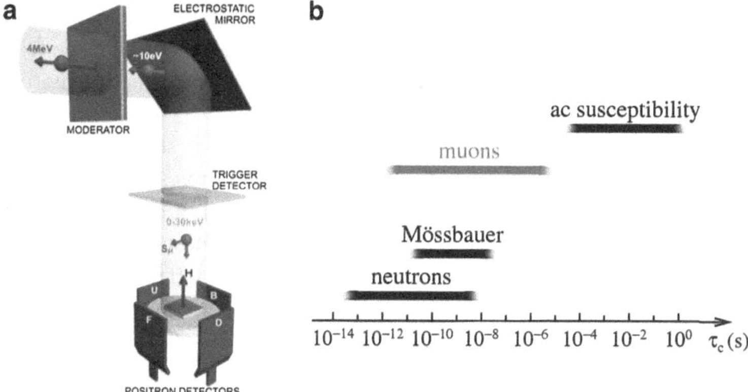

Fig. 3.5 **a** Schematic representation of a typical LE-μSR apparatus. Image courtesy A. V. Boris, Max Planck Institute for Solid-State Research, 70569 Stuttgart, Germany, Heisenbergstr. 1. **b** Time scales of dynamical processes accessible to various techniques. Image courtesy P. Dalmas de Réotier, CEA and University Joseph Fourier, Grenoble, France. Also available in black & white as Fig. 1.5 in Ref. [12]

entering a sample is dictated by the implantation depth required in the experiment. Current practical limitations typically allow one to achieve implantation depths from 10 to 300 nm, depending on the applied accelerating voltage and the sample density. This advantage of LE-μSR makes it one of the very few experimental techniques capable of investigating thin-film samples without any contaminating contribution from the substrate, on which the film has been deposited.

Since the only (albeit a very substantial) difference between conventional and LE-μSR is the energy of muons incident on a sample, the detection and data analysis procedures are the same. In a typical μSR experiment, millions of spin-polarized muons are implanted one at a time into a sample where they decay. The time evolution of the polarization of the muon ensemble $P(t)$ is followed by detecting the decay positron, which is emitted preferentially in the direction of the muon spin at the moment of the decay. The number of positrons detected by a counter as a function of time after the implantation reflects the time dependence of the muon spin polarization along the axis of observation defined by the detector:

$$N(t) = N_0 \exp(-t/\tau_\mu)(1 + A_0 P(t)) + N_{\mathrm{bg}},$$

where A_0 is the experimental decay asymmetry ($\approx$0.27 in our spectrometer), N_{bg} is the practically flat background of uncorrelated events, and τ_μ is the muon lifetime.

The quantity $P(t)$ contains all the information about the interaction of the muon spin with its local magnetic or spin environment and therefore provides information about the host material. Each muon stopped in the lattice (typically at an interstitial site) experiences the net effect of external and local magnetic fields and precesses with a characteristic Larmor frequency $\omega_{\mathrm{L}} = \gamma_\mu B$, where $\gamma_\mu/2\pi = 135.5$ MHz/T is the muon gyromagnetic ratio and B is the total magnetic field at the muon site. Thus μSR can measure local magnetic field distributions in the material. To see what kind of time dependence of the muon-spin polarization various magnetic environments produce, let us consider a few examples.

First of all, let us assume that the field at the muon site is stationary and that the sample under investigation is single-crystalline. In the simplest case of a dominant external magnetic field parallel to the original polarization of the muon spin (along z-axis) one obtains a time independent muon-spin polarization $P_z(t) = 1$ (all muon spins are parallel to the external field and thus do not precess). If the external magnetic field $\mathbf{B}_0$ is perpendicular to the original polarization of the muon spin, then all muons will precess with the same Larmor frequency $\omega_\mu = \gamma_\mu B_0$, where $B_0 = |\mathbf{B}_0|$ so that the muon-spin polarization becomes $P_z(t) = cos(\omega_\mu t)$. Now, if muons experience a distribution of magnetic fields $D(\mathbf{B})$ varying in both their magnitude and direction with respect to the original polarization of the muon spin, then the muon-spin polarization along z is obtained by averaging over all fields taking into account that the muon-spin component parallel to the magnetic field ($S\cos\theta$, where θ is the angle between the direction of the magnetic field and the muon spin S) is stationary (this stationary component will contribute $\cos\theta$ in the z-axis direction, that is, $S\cos^2\theta$), while the one perpendicular precesses with the corresponding Larmor frequency (and contributes $S\sin^2\theta$):

$$P_z(t) = \int \left(\cos^2\theta + \sin^2\theta\cos(\omega_\mu t)\right) D(\mathbf{B}) d^3B, \tag{3.6}$$

In a fully magnetically ordered *polycrystalline* material the magnitude of the magnetic field B_0 is the same but the direction is random, i.e. $D(\mathbf{B}) = \delta(B - B_0)/(4\pi B_0^2)$. Substituting this field distribution in the above equation one obtains

$$P_z(t) = \frac{1}{3} + \frac{2}{3}\cos(\omega_\mu t), \quad \omega_\mu = \gamma_\mu B_0. \tag{3.7}$$

If the distribution of magnetic fields at muon sites is Gaussian

$$D(\mathbf{B}) = \left(\frac{1}{\sqrt{2\pi}\Delta}\right)^3 \exp\left(-\frac{B^2}{2\Delta^2}\right), \tag{3.8}$$

as can be assumed for the magnetic field of ionic nuclei, then the muon-spin polarization in Eq. (3.6) reduces to the so-called Kubo-Toyabe function (shown in Fig. 3.6a)

$$P_z(t) = \frac{1}{3} + \frac{2}{3}\left(1 - \sigma^2 t^2\right)\exp\left(-\frac{\sigma^2 t^2}{2}\right), \tag{3.9}$$

where $\sigma^2 = \gamma_\mu^2\Delta^2$. In case an additional transverse external magnetic field B_{ext} is applied, the result changes to

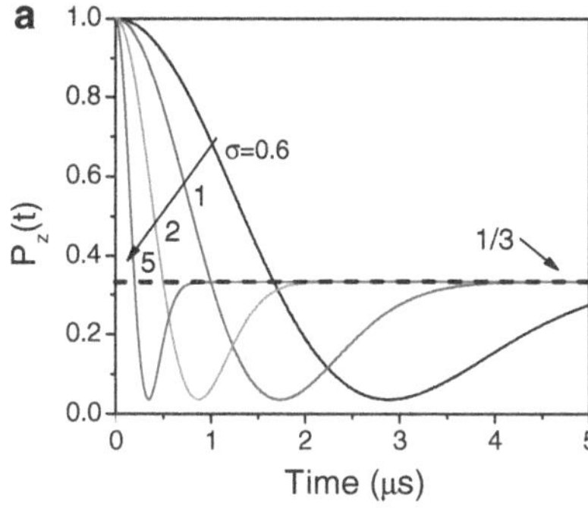

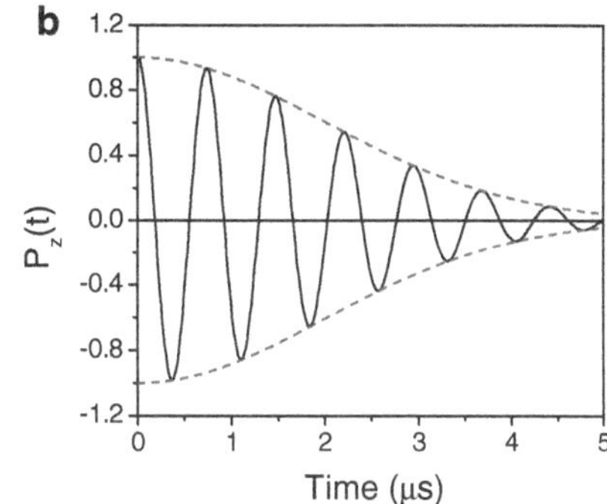

Fig. 3.6 **a** Muon-spin polarization along the direction of the original muon-spin polarization (at time zero) in the case of normally distributed local magnetic fields at the muon site for several values of the width of their statistical distribution σ in the absence of an external magnetic field. At large times the function approaches $1/3$, indicating that on average one third of all muons are unaffected by these local magnetic fields. **b** Muon-spin polarization along the direction of the original muon-spin polarization (at time zero) in the case of normally distributed local magnetic fields at the muon site and an additional external magnetic field of 100 G transverse to the direction of the original muon-spin polarization

$$P_z(t) = \exp\left(-\frac{\sigma^2 t^2}{2}\right)\cos(\omega_\mu t),$$

where $\omega_\mu = \gamma_\mu B_{\text{ext}}$. This function is plotted in Fig. 3.6b.

All above considerations are valid if magnetic fields at muon sites are stationary, i.e. change very slowly compared to the muon lifetime $\tau_\mu = 2.2$ μs. If, on the other hand, these magnetic fields are dynamic, all derivations should be modified accordingly. Such a situation can be realized not only due to the dynamic character of the fields themselves but also due to muon diffusion in the material. In the simplest case of the so-called motional-narrowing limit, when the dynamic change rate ν of the local fields at muon sites is much faster than σ in Eq. (3.9) (in the case of muon diffusion ν denotes their statistical hopping rate), the muon-spin polarization reduces to a particularly simple expression: $P_z = \exp(-\lambda t),\ \lambda = 2\sigma^2\tau$, where $\tau = 1/\nu$.

In the case of different magnetic environments, for instance, in a phase-separated material, $P(t)$ is a superposition of the corresponding signals. μSR is therefore able to detect and quantify different magnetic volume fractions in a sample.

In superconducting materials external magnetic fields get expelled from their bulk due to the screening supercurrents in the so-called Meissner effect [13]. In Type I superconductors external magnetic fields are entirely expelled up to a certain critical value H_c, which depends on the material and its temperature. Assuming that the penetration depth for the magnetic field into the sample is very small (high density of the superconducting condensate), that the sample is homogeneous, and neglecting all demagnetization effects due to the sample shape one can assume that in the superconducting state all muons in the material bulk sense no external field. The corresponding μSR signal is shown in Fig. 3.7a for a transverse external field of 100 G in the normal (black solid line) and superconducting (blue dashed line) state.

In Type II superconductors the situation becomes significantly more complex. There are now two critical fields, lower (H_{c1}) and upper (H_{c2}). Below H_{c1} the situation is identical to the case of Type I superconductors; above H_{c2} the material is in the normal state; for intermediate values the magnetic field partially penetrates through the sample through the so-called vortices (normal-state threads through a superconducting material along the direction of the external magnetic field), which arrange in a hexagonal (Abrikosov) lattice, as shown in the inset of Fig. 3.7b. Since the muon stopping sites have largely irregular positions with respect to vortices, the muons sample all values of the magnetic field in and around the vortices. This implies that the muon ensemble senses the field distribution of the vortex, which has a shape plotted in Fig. 3.7b. Such an inhomogeneous field distribution leads to gradual dephasing of precessing muon spins and thus to a decay of the detected spin polarization $P(t)$ in the superconducting state, as shown in Fig. 3.7a (blue open circles). In real materials some decay of the spin polarization in the presence of an external magnetic field exists already in the normal state due to local nuclear magnetic moments, providing a certain magnetic-field distribution at disordered muon stopping sites. In general, the vortex field profile in a particular material can be determined by carrying out the Fourier transform of $P(t)$ in the superconducting state. For most high-temperature cuprate and iron-based superconductors the lower critical field is on the order of 100 G—very small compared to the upper critical field, which ranges from several tesla to hundreds of tesla. For LE-μSR measurements the existence of the vortex lattice is further compounded by its dependence on the distance from the sample surface due to the finite penetration depth of external magnetic fields into the material under study. It leads to a funneling of the vortex-line magnetic-field profiles so as to make the field distribution progressively more homogeneous as one approaches

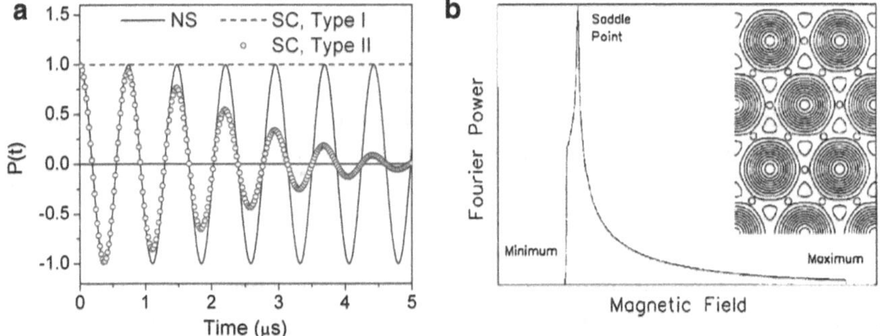

Fig. 3.7 **a** Time dependence of the muon-spin polarization $A_0 P(t)$ in a transverse external magnetic field of 100 G in the normal state (*black solid line*) and the superconducting state in the case of a Type I (*blue dashed line*) and Type II (*blue open circles*) superconductor. **b** Magnetic-field distribution in the hexagonal vortex lattice (see *inset*) of a Type II superconductor. Reprinted figure with permission from Ref. [14]. Copyright (2000) by the American Physical Society

the sample surface [15]. If one is interested in sample properties in the vicinity of its surface this effect must be taken into account explicitly.

3.4 Drude-Lorentz Model

One of the oldest and the most widely used analytical techniques in optical spectroscopy is dispersion analysis based on the so-called Drude-Lorentz model [16]. It allows one to extract basic characteristics of the intraband contribution to the optical conductivity (due to free charge carriers, itinerant response), the phonon contribution due to lattice vibrations, and the interband contribution due to particle-hole excitations between electronic bands. The model is based on the original description of free electrons in the framework of classical physics and was formulated by Drude in 1900 [17, 18]. The model treats all electrons in a metal as completely free, independent, and moving on the positive background of stationary positively charged ionic cores. The theory has seen tremendous success and provides a very good description of the optical and dc conductivity, Hall effect, and thermal conductivity of simple metals. Its applicability is, however, severely limited by the assumption of free and independent electrons, which immediately excludes a large fraction of metallic compounds in which electron-electron correlations dominate. In the latter case a full quantum-mechanical treatment of the problem is indispensable. In principle, based on ideas borrowed from the aforementioned quantum-mechanical description of correlated electrons, it is possible to generalize the Drude model in order for it to remain valid also in the presence of correlations by taking into account the retarded character of the resulting interaction. Such a generalization bears the name 'extended Drude model' and will be introduced below. Though helpful, it is largely an abstraction and should be considered as a convenient interface between experiment and theory, used to alleviate data analysis. As such it is only useful to the extent to which the underlying microscopic theory is known.

We will show below (see also Ref. [16]) that by introducing a restoring force acting on electrons from the positive background of ionic cores one can qualitatively describe interband particle-hole excitations in a material. This extension of the Drude theory was made in 1905 and is due to Hendrik Antoon Lorentz. The combined Drude-Lorentz theory has since been extensively augmented to incorporate various other results of the quantum theory (most notably in 1933 by Arnold Sommerfeld and Hans Bethe) but even in its simplest form it remains a useful tool of condensed-matter physics.

To derive his model, Drude made two assumptions: all itinerant electrons are independent and move freely except when encountering rigid immobile ionic cores, from which they scatter randomly. The classical equation of motion for each electron in an external monochromatic electric field $\mathbf{E}(\omega, \mathbf{t})$ then reads:

$$m\mathbf{x}'' = e\mathbf{E}(\omega, t) - m\gamma\mathbf{x}', \tag{3.10}$$

where the derivative is assumed with respect to time, m, e, and $\mathbf{x}(t)$ are the electron mass, charge, and coordinate vector, and γ is the scattering rate. Taking into account the time dependence of the external field $\mathbf{E}(\omega, \mathbf{t}) = \mathbf{E}_0 \exp^{-i\omega t}$ and looking only for a periodic solution of the form $\mathbf{x}(t) = \mathbf{x}_0 \exp^{-i\omega t}$ one arrives at

$$\mathbf{x}_0 = -\frac{e\mathbf{E}_0}{im\omega} \frac{1}{\gamma - i\omega}. \tag{3.11}$$

Using the classical expression for the current density $\mathbf{j} = ne\mathbf{v}$, where $\mathbf{j}$, n, and $\mathbf{v} = \mathbf{x}'$ are the electron current density, electron density, and speed, respectively, as well as its relation to the external electric field $\mathbf{j}(\omega, t) = \sigma(\omega)\mathbf{E}(\omega, t))$, where $\sigma(\omega)$ is optical conductivity [5], one obtains

$$\sigma(\omega) = \frac{ne^2}{m\gamma} \frac{1}{1 - i\omega\tau} = \frac{\sigma_0}{1 - i\omega\tau}, \tag{3.12}$$

where $\tau = 1/\gamma$ and $\sigma_0 = \sigma(0)$ is the dc conductivity. This simple but surprisingly widely applicable expression for the optical conductivity is one of the most important results of the Drude theory. Derived for the bare electron mass m of free, non-interacting electrons it remains valid in the framework of the Fermi liquid theory, in which rather than treating strongly interacting electrons one considers elementary excitations from the ground state of a correlated material, or quasiparticles. Although these quasiparticles describe collective excitations of many interacting electrons, they themselves often interact only weakly with their own like and thus, in certain cases, can be considered the independent counterparts of free electrons in the original Drude model with the exception of their mass being significantly modified by the strong interactions between the electrons. Thus the original Drude theory is equally applicable to weakly interacting *quasiparticles* of either electron or hole character in strongly correlated materials. It is this universality that has led to the tremendous success of the Drude model in condensed-matter research. For strongly interacting quasiparticles the Drude theory breaks down and the full quantum-mechanical treatment must be adopted to arrive at an adequate description of their optical properties. Even in the case of strongly interacting quasiparticles the optical conductivity retains its Drude-like form of Eq. (3.12) but with a frequency-dependent effective quasiparticle mass $m_{\text{eff}}(\omega)$ and scattering rate $\gamma(\omega)$:

$$\sigma(\omega) = \frac{\omega_{\text{pl}}^2}{4\pi} \frac{1}{\gamma(\omega) - i\omega m_{\text{eff}}(\omega)/m}.$$

This modified description of the itinerant charge carrier response bears the name of 'extended Drude model'. Although physically meaningful, describing the mass and the scattering rate of quasiparticles when driven by a monochromatic electric field at frequency ω, this simple expression incorporates two independent functions of frequency $m_{\text{eff}}(\omega)$ and $\gamma(\omega)$ and is thus a mere analytical transformation of the optical conductivity. It implies that any optical response, including the contribution of physical phenomena unrelated to itinerant quasiparticles [such as lattice vibrations (phonons) or interband particle-hole excitations], can be converted into $m_{\text{eff}}(\omega)$, and $\gamma(\omega)$. In such a case, however, these quantities have no direct physical meaning. Therefore care must be taken to ascertain that only the contribution of quasiparticles to the overall optical response is used for the analysis in the extended Drude model.

In 1905 Hendrik Antoon Lorentz extended the Drude model by considering a restoring force exerted on free charge carriers by the positive ions of the lattice. By assuming small electronic displacements this force can be expanded in powers of $\mathbf{x}_0$, with only the first-order term $-\kappa\mathbf{x}_0$ retained (it acts to prevent further displacement, hence the sign). Introducing this force on the right-hand side of Eq. (3.10) and using the standard definition of the resonance frequency $\omega_0^2 = \kappa/m$, one arrives at

$$m\mathbf{x}'' = e\mathbf{E}(\omega, t) - m\gamma\mathbf{x}' - m\omega_0^2\mathbf{x}.$$

Assuming the external electric field to be monochromatic and looking only for a periodic solution, by analogy with the above derivation of the Drude model, one obtains the following electronic displacement amplitude:

$$\mathbf{x}_0 = -\frac{e}{m}\mathbf{E}_0\frac{1}{\left(\omega^2 - \omega_0^2\right) + i\gamma\omega}. \tag{3.13}$$

This functional dependence, named after Lorentz, has been found to provide a good description of the optical response due to lattice vibrations (phonons) as well as, in some cases, particle-hole excitations. Based on the above expression for the atomic displacement one can obtain the corresponding optical conductivity by following the same steps as in the derivation of the Drude counterpart.

The simple derivation of the Drude-Lorentz model presented above can be used to establish a connection between the optical conductivity and the dielectric function of a material. Rather than describing the current induced in the material by an external electric field, as is the case for the optical conductivity, the dielectric function connects, via the electric susceptibility, the induced macroscopic polarization $\mathbf{P}(\mathbf{E})$ to the external electric field. In a linear material, one can expand the polarization in powers of the field and preserve only the first-order term: $\mathbf{P}(\omega, \mathbf{t}) = \chi(\omega)\mathbf{E}(\omega, \mathbf{t})$, where $\chi = \chi_{ij}$, $i, j = x, y, z$ is the electric susceptibility and is a tensor quantity, transforming the electric-field vector into the polarization vector. For an isotropic medium $\chi_{ij} = \chi\delta_{ij}$ and only one independent complex component remains. The macroscopic polarization can be calculated as the dipole moment per unit volume. In the simplest case of the free-electron model the amplitude $\mathbf{x}_0$ obtained above describes the displacement of the negatively charged itinerant carriers with respect to the positive background, i.e. the formation of a dipole moment $\mathbf{p}_0 = e\mathbf{x}_0$. Then,

$$\mathbf{P}(\omega, t) = \frac{Ne\mathbf{x}_0}{V} = -\frac{ne^2}{mi\omega}\frac{1}{\gamma - i\omega}\mathbf{E}_0 = \chi\mathbf{E}_0, \tag{3.14}$$

where N is the number of electrons in the volume V. By comparing the above equation to Eq. (3.12) one infers that $\sigma = -i\omega\chi$. The dielectric function is defined as $\varepsilon = 1 + 4\pi\chi$, therefore

$$\varepsilon(\omega) = 1 + \frac{4\pi i}{\omega}\sigma(\omega). \tag{3.15}$$

The above derivations have been carried out in CGS units. To convert to other units one can use the standard conversion rules [5]. For instance, if ω is measured in cm^{-1} and $\sigma(\omega)$—in $\Omega^{-1}\,\mathrm{cm}^{-1}$, then the factor 4π in Eq. (3.15) should be replaced by 60.

Due to the above considerations, the Drude and Lorentz contributions to the dielectric function of a medium (corresponding $4\pi\chi$ terms) have the following form:

$$\begin{aligned}\varepsilon_{\mathrm{Drude}}(\omega) &= -\frac{\omega_{\mathrm{pl}}^2}{\omega^2 + i\gamma\omega},\\ \varepsilon_{\mathrm{Lorentz}}(\omega) &= \frac{\Delta\varepsilon_0\omega_0^2}{\left(\omega_0^2 - \omega^2\right) - i\gamma\omega},\end{aligned} \tag{3.16}$$

where the plasma frequency of free charge carriers ω_{pl}^2 and the oscillator strength $\Delta\varepsilon_0\omega_0^2$ of the process described by the Lorentzian term have been introduced for $4\pi ne^2/m$ in the Drude and Lorentz terms, respectively. The reason for choosing different notation for nominally the same microscopic quantity lies in the fact that the above expressions for the dielectric function have been found to describe the optical response due to various physical processes, even those outside the scope of the Drude-Lorentz model. In the case of the Lorentz term, $\Delta\varepsilon_0$ gives the contribution of the respective Lorentzian oscillator to the zero-frequency real part of the dielectric function: $\varepsilon_1(\omega \to 0)$. In the case of a purely itinerant response, on the other hand,

$$\varepsilon(\omega) = 1 - \frac{\omega_{\mathrm{pl}}^2}{\omega^2 + \gamma^2} + i\frac{\omega_{\mathrm{pl}}^2\gamma}{\omega^3 + \gamma^2\omega}. \tag{3.17}$$

Therefore, in the absence of scattering ($\gamma = 0$) the real part of the dielectric function becomes zero at $\omega = \omega_{\mathrm{pl}}$. At this frequency all itinerant electrons oscillate collectively with respect to the positive ionic background as they would in plasma, hence the name 'plasma frequency'. If scattering is not negligible and if there is a significant contribution to the real part of the dielectric function from excitations at higher energies, which would approximately result in the addition of $\Delta\varepsilon_{\mathrm{tot}} = \sum_j \Delta\varepsilon_{0,j}$ to the above value of $\varepsilon(\omega)$, the value of the actual plasma frequency changes from its formal definition to

$$\Omega_{\mathrm{pl}}^2 = \frac{\omega_{\mathrm{pl}}^2}{\Delta\varepsilon_{\mathrm{tot}}}\sqrt{1 - \gamma^2\Delta\varepsilon_{\mathrm{tot}}/\omega_{\mathrm{pl}}^2}. \tag{3.18}$$

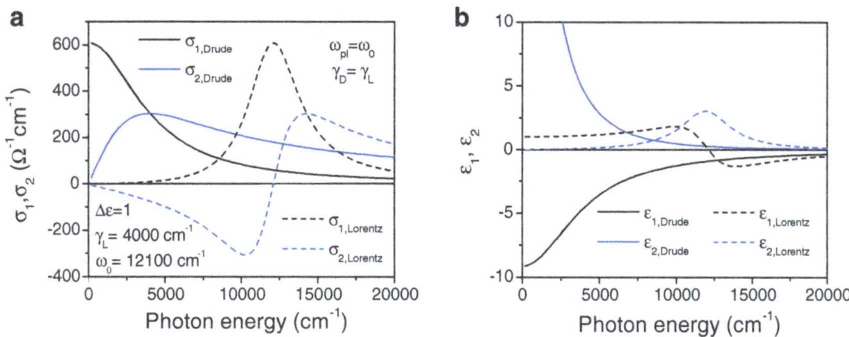

Fig. 3.8 Real and imaginary part of the optical conductivity (**a**) and the dielectric function (**b**) of the Drude and Lorentz terms given by Eqs. (3.15, 3.16) (without the unity contribution to the real part of the dielectric function)

Figure 3.8 illustrates the functional dependence of the optical conductivity and the dielectric function for both kinds of response for the same values of common parameters.

The most fundamental laws of physics have always been those based on symmetries. If a physical system is symmetric with respect to a certain transformation then all its observable properties must be conserved under the application of this transformation. In this way the translation symmetry in real space results in the conservation of total momentum, that in time results in the conservation of total energy. These are the symmetries of classical physics, whose discovery probably traces back to the work of Newton. The development of quantum physics led to the discovery of other, more subtle symmetries. One of such symmetries is the invariance of physical phenomena with respect to the unitary transformation of the electromagnetic field, which results in the conservation of charge. For an isolated material in a condensed state it can be reformulated as the conservation of the total number of electrons. In terms of the optical properties of a system, this symmetry gives rise to the so-called f-sum rule [1], which stipulates that the total area under the $\sigma_1(\omega)$ curve of any material is equal to $\pi ne^2/(2m)$, where m is the *bare* electron mass, and is conserved. The area under a conductivity curve is called spectral weight, so that the total spectral weight of any system is always conserved. This fundamental conservation law often proves useful in studying the changes of optical properties across phase transitions, as will be shown in the following section and in Sect. 4.1. Figure 3.8a illustrates this sum rule: although dramatically different in their functional form, the Drude and Lorentz terms have exactly the same spectral weight because their integrals are equal to $\omega_{\mathrm{pl}}^2/8$ and $\Delta\varepsilon_0\omega_0^2/8$ in CGS units (in the units of Fig. 3.8a the denominator 8 should be replaced with 38), respectively, and in the case presented $\omega_{\mathrm{pl}}^2 = \Delta\varepsilon_0\omega_0^2$.

The f-sum rule has also turned out to be of much use in the theory of superconductivity. Already the first optical studies of superconducting materials [19] revealed that the area under the conductivity curve is significantly reduced in the superconducting state compared to the normal state at a slightly higher temperature. Since the total spectral weight is always conserved, this observation implies that the missing area must be transferred to some other spectral range. These arguments led to the discovery of the so-called Ferrell-Glover-Tinkham rule, according to which the missing area between the normal and superconducting state is transferred to the dc optical response at zero frequency and enables nondissipative electron transport in the superconducting state. This sum rule is a powerful spectroscopic tool widely used in the investigation of superconducting materials. It is particularly informative in the case of spectroscopic ellipsometry, as will be discussed in the next section.

3.5 Kramers-Kronig Consistency Analysis

In the previous section it has already been mentioned that the KK relations (3.2) are widely used in various spectroscopic techniques. They are of particular importance for reflectivity measurements, in which the use of the mutual dependence of the real and imaginary part of the dielectric function is the only way to extract them from a single function of frequency—relative intensity of reflected light. Spectroscopic ellipsometry, on the other hand, boasts the ability to measure both the real and the imaginary part of the dielectric function simultaneously and independently, without any reference required. It means that in this case one can fully dispense with the KK relations. However, it is precisely because they are not required for extraction of the optical response from raw data obtained with spectroscopic ellipsometry that the KK relations become a powerful tool of data analysis, allowing one to obtain information about the optical properties *outside* the experimentally accessible range. Such information can be inferred based on a consistency analysis of the directly measured real or imaginary part of the dielectric function or the optical conductivity and its counterpart resulting from the transformation of its other part using the KK relations, after it has been extrapolated to cover the entire spectral range $\omega \in [0, \infty)$, as required by Eqs. (3.2). Although such low-energy extrapolation cannot be uniquely determined by the KK consistency analysis, the *spectral weight* contained in the extrapolation region can. It appears to be of much importance in the analysis of superconducting properties due the spectral weight transfer from finite energies in the far-infrared spectral range to the δ-function at zero frequency, as already mentioned above. Since it is impossible to access the zero-frequency response with a spectroscopic technique, the KK consistency analysis can shed light onto such otherwise moot questions as whether the missing area in the optical conductivity is indeed fully transferred to the δ-function at zero frequency in the superconducting state (in other words, whether the Ferrell-Glover-Tinkham sum rule is satisfied) or if processes at much higher energies than the superconducting energy gap contribute to the formation of the superconducting condensate, as can be expected from some exotic mechanisms of superconductivity [20, 21]. In this section we will illustrate the procedure of the KK consistency analysis and assess what kind of information can be reliably inferred from it.

Before examining the application of the KK consistency analysis to the modification of optical properties at a superconducting transition, let us first examine the procedure in the simplest case of a purely itinerant optical response given by the Drude term, as shown in Fig. 3.9 (black line). We assume the lowest experimentally accessible frequency to be 50 cm^{-1}. Since the exact shape of the optical response below this frequency is unknown, one must resort to a certain extrapolation of the experimentally obtained spectrum. To simulate this situation and study its effects on the inferred properties we apply the KK transformation (3.2) to the real part of the optical conductivity of the same Drude response slightly modified in the extrapolation region (blue circles in Fig. 3.9a). The resulting real part of the dielectric function (blue circles in Fig. 3.9b) is compared to that of the unchanged Drude term in Fig. 3.9b. It is clear that although the changes in the real part of the optical conductivity are constrained to the spectral range below 50 cm^{-1}, those in the real part of the dielectric function extend to much higher frequencies. Given a typical experimental noise level of $\delta\varepsilon_1 \approx 30$ in this spectral range for such metallic samples (see, e.g. Ref. [22]), the deviation of $\varepsilon_1(\omega)$ obtained from incorrectly extrapolated $\sigma_1(\omega)$ can still be resolved at frequencies as high as 300 cm^{-1}. Thus the uncertainty of the optical response in the extrapolation region strongly affects the extracted optical properties at much higher frequencies. This is a simple consequence of the fact that the application of the KK relations requires the knowledge of one of the components of the optical response in the *entire* spectral range. This argument demonstrates the strong advantage of ellipsometric measurements, in which both the real and the imaginary part of the optical response are obtained independently and thus are unaffected by the lack of information in the experimentally inaccessible low-energy spectral range. It is noteworthy that the difference of the spectral weight of the two spectra shown in Fig. 3.9a constitutes only 4 % of the total spectral weight of the Drude term. Such a

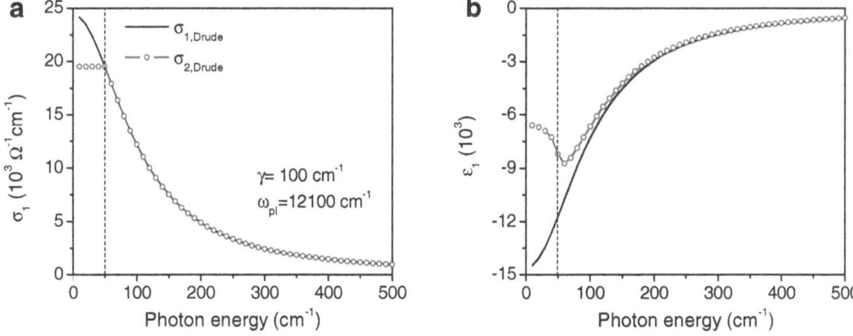

Fig. 3.9 Real part of the optical conductivity (**a**) and the dielectric function (**b**) of the Drude term given by Eqs. (3.15, 3.16) (without the unity contribution to the real part of the dielectric function, *black lines*). *Blue circles* in **a** show the optical conductivity of the same Drude term modified in the 'extrapolation' spectral range below 50 cm^{-1} and those in **b** display the corresponding real part of the dielectric function obtained via the KK transformation (3.2). *Dashed vertical lines* mark the upper limit of the extrapolation spectral range

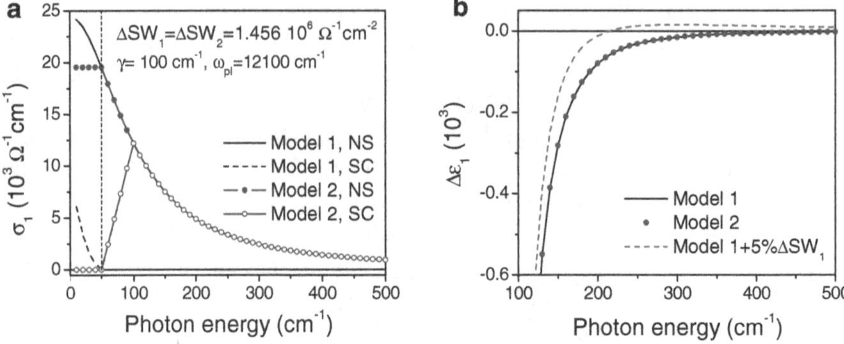

Fig. 3.10 **a** Real part of the optical conductivity for two models of the Drude-like optical conductivity (*black lines* and *blue symbols*) in the normal and superconducting state (*solid line* and *filled circles*—normal state, *dashed line* and *open circles* — superconducting state). The missing area between the spectra of the optical conductivity in the normal and superconducting state ($\Delta SW_{1,2}$ for models 1 and 2, respectively) is transferred into the δ-function at zero frequency. The models are chosen in such a way that the missing area is the same: $\Delta SW_1 = \Delta SW_2 = 1.456 \times 10^6\ \Omega^{-1}\text{cm}^{-2}$. Above 50 cm^{-1} the respective spectra in the normal and the superconducting state coincide. Above 100 cm^{-1} all spectra coincide. **b** Difference spectra of the real part of the dielectric function obtained using Eq. (3.19) given the real part of the optical conductivity from the two models shown in **a**. The two results are indistinguishable on the scale of the figure (the difference between the two is ≈ 10 at 100 cm^{-1} and less than 1 above 200 cm^{-1}). In addition, *the red dashed line* shows the difference spectrum of the real part of the dielectric function for model 1 with additional spectral weight (5 % of ΔSW_1) transferred to the δ-function at zero frequency from very high energies

KK consistency analysis beyond the extrapolation spectral range is mostly sensitive to the difference in the spectral weight between different kinds of extrapolation and not to their exact shape in this region. This issue will be analyzed in more detail below.

In the case of normal-state spectral properties at a given temperature, performing reflectivity measurements down to lower frequencies improves the accuracy of the extraction of the optical response. Contemporary state-of-the-art reflectivity spectrometers using far-infrared radiation of bright synchrotron sources can carry out reliable measurements down to 20 cm^{-1} under optimal conditions. In the case of a superconducting transition, however, even such measurements leave significant room for uncertainty. This becomes clear if one remembers that, according to the Ferrell-Glover-Tinkham rule, the missing area in the real part of the optical conductivity is transferred into the δ-function at zero frequency, describing the coherent response of the superconducting condensate. While irrelevant at finite frequencies for the real part of the optical conductivity, this singular part of the response dramatically changes the imaginary part of the optical conductivity, as will be shown in what follows. Since the exact spectral weight of the extrapolation region is unknown, this gives rise to a residual uncertainty in the inferred optical properties. Let us now consider it in more detail.

Due to the Ferrell-Glover-Tinkham rule the KK relations in the superconducting state must be rewritten to separate the singular optical response at zero frequency from the remaining nonsingular part. In what follows we concentrate on the KK relation expressing $\varepsilon_1(\omega) - 1$ via an integral of $\varepsilon_2(\omega)$ and choose to work with $\sigma_1(\omega) = \omega\varepsilon_2(\omega)/60$ (in SI units) over $\varepsilon_2(\omega)$. Writing out in the superconducting state $\sigma_{\text{total}}(\omega) = SW\delta(\omega) + \sigma(\omega > 0)$, where SW denotes the spectral weight transferred to the δ-function from higher frequencies, and remembering that $\int f(\Omega)\delta(\Omega - \omega)d\Omega = f(\omega)$, the first of the KK relations (3.2) can be rewritten in SI units as:

$$\varepsilon_1(\omega) - 1 = 38\,\mathscr{P}\int_0^\infty \left[-\frac{SW}{\omega^2} + \frac{\sigma_1(\Omega)}{\Omega^2 - \omega^2}\right] d\Omega \tag{3.19}$$

(in CGS units the prefactor 38 should be replaced by 8). If the superconducting pairing interaction is confined to small energies than the Ferrell-Glover-Tinkham rule is strictly fulfilled and SW is composed of the far-infrared missing area only. In some cases, however, higher-energy processes might contribute to the formation of Cooper pairs, which leads to additional spectral weight transferred to the δ-function at zero frequency from energies, much higher than those the Ferrel-Glover-Tinkham rule originally concerned. In such situations, reflectivity measurements and data analysis would be unable to answer the question of whether high-energy processes contribute to SW, whereas the KK consistency analysis of ellipsometric data can easily address this issue. To illustrate this statement one can compare the following three model cases (shown in Fig. 3.10):

- Model 1: the normal state $\sigma_1(\omega)$ is given by the Drude term with the same parameters as in Fig. 3.9a (black solid line in both Fig. 3.9a and Fig. 3.10a), while that in the superconducting state (black dashed line) shows superconductivity-induced suppression with residual absorption at low energies and a total missing area $\Delta\text{SW}_1 = \int \left(\sigma_1^{\text{SC}}(\Omega) - \sigma_1^{\text{NS}}(\Omega)\right) d\Omega = 1.456 \times 10^6\ \Omega^{-1}\text{cm}^{-2}$.
- Model 2: the normal state is the same as the truncated Drude term in Fig. 3.9a (blue filled circles in Fig. 3.10a), while the superconducting state (blue open circles in Fig. 3.10a) shows zero $\sigma_1(\omega)$ below 100 cm^{-1} with precisely the same missing area as in Model 1, albeit a quite different spectral shape.
- Model 3: the finite-frequency optical response is identical to Model 1 in both the normal and superconducting state but additional spectral weight (5 % of ΔSW_1) is transferred to the δ-function from very high energies, such that the direct change in the real part of the dielectric function due to the suppression of a high energy transition is negligible. The latter condition can always be satisfied by choosing the transition to be located at high enough energies. If one approximates this transition by the Lorentzian term from Eq. (3.16) than its total spectral weight is given by $\text{SW}_\text{L} = \Delta\varepsilon_0\omega_0^2/8$, as it has been shown above. Given that $\omega_0^2 \gg \Delta\varepsilon_0$, even the complete loss loss of the spectral weight in this transition would lead to a change in the real part of the dielectric function of $\Delta\varepsilon_0 = 8\text{SW}_\text{L}/\omega_0^2$, which can be made arbitrarily small.

One can now investigate the real part of the dielectric function for these three models obtained using the KK relation for the superconducting state, Eq. (3.19). Since only the difference properties between the normal and superconducting state are usually of interest and because working with difference spectra eliminates all systematic temperature-independent contaminating contributions, we will only study $\Delta\varepsilon_1(\omega) = \varepsilon_1^{\mathrm{NS}}(\omega) - \varepsilon_1^{\mathrm{SC}}(\omega)$. Figure 3.10b shows the result of the application of the KK transformation (3.19) to the three models introduced above. It is immediately apparent that $\Delta\varepsilon_1(\omega)$ for Models 1 and 2 are indistinguishable on the scale of the figure with their difference being $\approx$10 at 100 cm^{-1} and less than 1 above 200 cm^{-1}. This result illustrates that the exact *shape* of the extrapolation used in the KK consistency analysis has very little effect on the inferred difference spectra. The very important distinction from the use of the KK relations in reflectivity measurements consists in the fact that here this transformation is not used to *extract* the optical properties from intensity measurements but rather to determine the precise amount of the spectral weight contained in the extrapolation region (including the δ-function at zero frequency) based on the independently obtained real and imaginary part of the dielectric function. This point is further illustrated by the fact that although the shape of the extrapolation used for $\sigma_1(\omega)$ has a very small effect on $\Delta\varepsilon_1(\omega)$, even a very small variation in the total *spectral weight* in the extrapolation region (5 % in Model 3) leads to much larger changes that can be easily detected in routine ellipsometric measurements (red dashed line in Fig. 3.10b). For instance, based on the difference between the red dashed and the black solid line in Fig. 3.10b and a typical noise level in the real part of the dielectric function on the order of 30 at 200 cm^{-1} already mentioned above, one can confidently detect additional spectral weight as small as 2 % of the total spectral weight at zero frequency. Upon carrying out more detailed ellipsometric measurements with much larger statistical averaging it is possible to reduce the noise level in $\Delta\varepsilon_1(\omega)$ by an order of magnitude and thus boost the sensitivity into the sub-percent range.

To conclude this section we examine the weak dependence of the KK consistency analysis on the shape of the low-energy extrapolation more closely. The maximum uncertainty introduced by the unknown extrapolation shape of $\Delta\sigma_1(\omega)$ can be calculated as the difference between two extreme configurations: all $\Delta\mathrm{SW}(\omega_0)$ at $\omega = 0$ and $\omega = \omega_0$:

$$\begin{aligned}\delta\Delta\varepsilon_1^{(1)}(\omega) &= 8\left[-\frac{\Delta\mathrm{SW}(\omega_0)}{\omega^2} - \frac{\Delta\mathrm{SW}(\omega_0)}{\omega_0^2 - \omega^2}\right] \\ &= 8\frac{\Delta\mathrm{SW}(\omega_0)}{\omega^2}\frac{\omega_0^2}{\omega_0^2 - \omega^2} \longrightarrow 8\frac{\Delta\mathrm{SW}(\omega_0)}{\omega^2}\frac{\omega_0^2}{\omega^2},\end{aligned}$$

when $\omega \gg \omega_0$. On the other hand, the accuracy to which the *spectral weight* is determined at the same energy is given by

$$\left|\delta\Delta\varepsilon_1^{(2)}(\omega)\right| = \left|8\int_{0+}^{\infty}\frac{\delta\Delta\sigma_1(x)}{x^2 - \omega^2}dx\right| \geq \left|8\frac{\delta(\Delta\mathrm{SW}(\omega_0))}{\omega^2}\right|.$$

The relative effect of the shape change over the magnitude change of the spectral weight in the extrapolation region is then $|\delta\Delta\varepsilon_1^{(1)}(\omega)/\delta\Delta\varepsilon_1^{(2)}(\omega)| \longrightarrow (\omega_0/\omega)^2$

and rapidly goes to zero for $\omega \gg \omega_0$. For instance, taking $\omega_0 = 12$ meV and $\omega = 250$ meV one gets the upper boundary for the relative shape uncertainty on the order of 0.2 %. Thus the effect is negligible already at rather low frequencies.

3.6 Eliashberg Theory of Superconductivity

After the discovery of superconductivity by Onnes in 1911 this phenomenon and its origin captivated the minds of the greatest theoreticians of the era. Richard Feynman, for instance, managed to explain the macroscopic coherence of superfluid helium and tried to find the key to the mystery of superconductivity but the solution eluded him. It was not until almost 50 years after the discovery of superconductivity that the first successful microscopic theory of the phenomenon was introduced by Bardeen, Cooper, and Schrieffer. Although this theory made several crude approximations it was nevertheless capable of describing the properties of most weakly coupled superconductors known at the time. Subsequent work of Gor'kov [23], Midgal [24], and Eliashberg [25] led to a more accurate and consistent description of the electron-phonon as well as electron-electron Coulomb interaction in superconductors and succeeded in addressing the departures from the BCS theory of superconductivity found in some materials, such as Pb, which were thus realized to be strongly coupled superconductors and therefore lay beyond the applicability range of the BCS theory. This development over the weak-coupling BCS theory, which received the name of the Eliashberg theory of superconductivity, after Gerasim Matveevich Eliashberg, who first introduced a consistent treatment of the retarded, frequency-dependent electron-phonon interaction into the theory of superconductivity, has since been successfully used to describe the superconducting properties of a large number of superconducting materials and to this day remains the most complete theory of superconductivity. An instructive historical account of the development of the BCS and Eliashberg theory of superconductivity can be found in the Nobel lectures of Bardeen, Cooper, and Schrieffer.

The main improvement of the Eliashberg theory over the BCS theory of superconductivity is in the consistent treatment of the electron-phonon interaction not merely as a source of attractive electron-electron interaction but in its entirety, including the kinetic effect of such interaction on electrons, that is, the modification of their energy spectrum: $\varepsilon(\mathbf{k}) = \varepsilon_0(\mathbf{k}) + \Sigma(\omega_{\mathbf{k}})$. This additional contribution to the energy of independent itinerant electrons is called self-energy because it describes the modification of the intrinsic electron kinetic energy due to virtual processes of emission and reabsorption of a phonon, i.e. even at absolute zero (when no real phonons exist and thus no electron-phonon scattering takes place). In this respect it is equivalent to the phenomenon of the increase of the free-electron mass due to its interaction with the electromagnetic vacuum (virtual emission and reabsorption of a photon). Any modification of the electron energy spectrum leads, of course, to a corresponding change in the electron mass because the latter is defined as

$$\frac{1}{m_\alpha(\omega_{\mathbf{k}})} = \frac{1}{\hbar^2}\frac{d^2\varepsilon(\mathbf{k})}{d^2k_\alpha}, \quad \alpha = x, y, z. \tag{3.20}$$

In an isotropic material $m_\alpha(\omega_{\mathbf{k}}) = m(\omega_{\mathbf{k}})$. The ratio of this and the bare electron mass is called mass renormalization and can be written as $m(\omega_{\mathbf{k}})/m_0 = 1 + \lambda$, where λ characterizes the coupling strength between electrons and phonons. Many other properties of itinerant electrons are renormalized as well [26]. In addition, different experimental techniques are often sensitive to different manifestations of the same physical parameter, such as mass, and the functional dependence of their renormalization on the coupling strength can be quite different.

When a physical quantity determined with a certain experimental technique is independent of the reciprocal-space structure and is only sensitive to the overall energy spectrum of interactions, one can average the corresponding interaction strength over the reciprocal space at a given energy. In the case of the electron-phonon interaction this average is denoted as $\alpha^2(\omega)F(\omega)$ and, in the simplest case, is a product of the electron-phonon interaction constant at frequency ω averaged over the reciprocal space and the polarizations of phonons $\alpha^2(\omega)$ (which determines how strongly electrons are coupled to phonons) and the phonon DOS $F(\omega)$ (which describes how many phonon states are available at this energy to interact with in the first place). In terms of this function $\alpha^2(\omega)F(\omega)$ (which is often called the Eliashberg function) the parameter λ can be expressed as:

$$\lambda = 2\int_0^\infty \frac{\alpha^2(\omega)F(\omega)}{\omega}d\omega. \tag{3.21}$$

Migdal was the first one to realize the importance of the consistent treatment of the electron-phonon interaction to describe the normal-state properties of metals [24]. Eliashberg then used these ideas and the formalism for calculating superconducting properties developed by Gor'kov [23] to derive a set of self-consistent equations for the superconducting order parameter $\Delta(\omega)$ and the quasiparticle renormalization $Z(\omega)$ in terms of the function $\alpha^2(\omega)F(\omega)$ [25]. Often formulated on the imaginary axis for the so-called 'thermodynamic time' $0 \le \tau \le 1/T$ and fermionic Matsubara frequencies $\omega_{\mathrm{n}} = \pi T(2n-1)$, where T is the temperature and $n = 0, \pm1, \pm2, \ldots$, these equations have the following form [27]:

$$Z(i\omega_n) = 1 + \frac{\pi T}{\omega_n}\sum_m \lambda(i\omega_m - i\omega_n)\frac{\omega_m}{\sqrt{\omega_m^2 + \Delta^2(i\omega_m)}}, \tag{3.22}$$

$$\Delta(i\omega_n)Z(i\omega_n) = \pi T\sum_m \left[\lambda(i\omega_m - i\omega_n) - \mu^*(\omega_{\mathrm{c}})\theta(\omega_{\mathrm{c}} - |\omega_m|)\right] \times \frac{\Delta(i\omega_m)}{\sqrt{\omega_m^2 + \Delta^2(i\omega_m)}},$$

where $\lambda(i\omega_m - i\omega_n)$ is defined in terms of the Eliashberg function as

$$\lambda(i\omega_m - i\omega_n) = 2\int_0^\infty \frac{\Omega\alpha^2(\Omega)F(\Omega)}{\Omega^2 + (\omega_m - \omega_n)^2}d\Omega, \tag{3.23}$$

and $\mu^*(\omega_c)$ is the Coulomb pseudopotential (accounting for the electron-electron Coulomb interaction) and comes with a cutoff ω_c lest the sum diverge. The value of this pseudopotential is rather difficult to determine from first principles so it is usually considered a free parameter. The single-band (or, equivalently, defined on a single sheet of the Fermi surface) Eliashberg equations above can be easily generalized to an arbitrary number of bands s (sheets of the Fermi surface), in which case there appear s order-parameter and renormalization functions, the coupling function becomes an $s \times s$ matrix $\lambda_{ij}(i\omega_m - i\omega_n)$, $i, j = \overline{1, s}$, and another summation is introduced into both equations with respect to the band index.

Once the Eliashberg equations have been solved on the imaginary axis, many frequency-independent material properties such as thermodynamic functions (free energy, specific heat) as well as the superconducting transition temperature can be obtained directly from $\Delta(i\omega_n)$ and $Z(i\omega_n)$. If one is interested in spectral properties such as the optical conductivity or dielectric function, tunneling DOS etc., the solution must be analytically continued to the real axis to obtain the real-frequency spectral functions $\Delta(\omega)$ and $Z(\omega)$ [27]. Once this has been done various spectral properties can be evaluated. For instance, the tunneling DOS in the superconducting state can be obtained from the real-frequency solution of the Eliashberg equations as follows:

$$N_T(\omega) = \frac{\partial}{\partial \omega} \int N(E)\,[f(E+\omega) - f(E)]\,dE, \tag{3.24}$$

where $f(E)$ is the Fermi-Dirac statistical distribution and $N(E)$ is the DOS in the superconducting state obtained from the solution of the Eliashberg equations via

$$N(\omega) = \sum_{i=1}^{s} N_i(0) \mathrm{Re}\left[\frac{\omega}{\sqrt{\omega^2 + \Delta_i^2(\omega + i\delta)}}\right], \tag{3.25}$$

where s is the number of bands and $N_i(0)$ is the ith-band's DOS in the normal state at the Fermi surface. The calculation of the optical response requires a little more effort, as well as the knowledge of several additional properties of the material under study, such as the impurity scattering rate in each band, those between the bands, and the plasma frequencies of the bands. In addition, the Eliashberg function $\alpha^2(\omega)F(\omega)$ must be replaced by its 'transport' counterpart, which takes into account the Fermi velocities of electrons and holes when averaging over the reciprocal space. In general, the anisotropic optical conductivity $\sigma_{\mu\nu}(\mathbf{k}, \omega)$, $\mu, \nu = x, y, z$ can be obtained from the paramagnetic response function $K_{\mu\nu}(\mathbf{k}, \omega)$ via

$$\sigma_{\mu\nu}(\mathbf{k}, \omega) = \frac{1}{4\pi i \omega} K_{\mu\nu}(\mathbf{k}, \omega). \tag{3.26}$$

This paramagnetic response function $K_{\mu\nu}(\mathbf{k}, \omega)$ is equivalent, for positive frequencies, to the Fourier transform of the time-ordered current-current correlation function

$$P_{\mu\nu}(x, x') = -4\pi i \langle T j_\mu(x) j_\nu(x') \rangle, \tag{3.27}$$

where T is the time ordering operator, $j_\mu(x)$ is the current operator at a space-time point x. For a very detailed account of the electromagnetic properties of superconductors see Refs. [28, 29].

Since the details of the Eliashberg theory and the electromagnetic properties of superconductors have been reviewed in numerous works, some of which have been cited above, we will not repeat them here. Rather, it appears more instructive to put the development of the microscopic theory of superconductivity and the understanding of its applicability to various types of superconductors into perspective.

As it has already been outlined in the introduction, the BCS theory of superconductivity was the first microscopic theory of the phenomenon. It succeeded in the description of many superconducting properties such as the existence of the superconducting energy gap in the DOS and the excitation spectrum of quasiparticles and its dependence on temperature, the existence of an anomaly in the specific heat of superconductors, and many others. However, from the very beginning the theory was formulated as a *single-band weak-coupling theory* of superconductivity in that the Fermi surface was assumed to have only one sheet and the electron-phonon interaction to be weak so that the renormalization of the electron energy spectrum could be neglected. The first limitation was overcome very shortly when the BCS theory was extended to the case of several bands [30]. One year later the second limitation was gotten rid of with the formulation of the Eliashberg theory of superconductivity for arbitrary electron-phonon interaction strength. The single-band BCS theory can be easily shown to follow from the Eliashberg theory when the original assumption of the former are made in the latter (in the so-called two-square-well model, in which $\lambda(i\omega_m - i\omega_n) = \lambda$, when $|\omega_{m,n}| < \omega_c$ and 0 otherwise and with the coupling assumed to be weak ($\lambda \ll 1$)) [27]. Quite interestingly, and this fact was not uncovered until the next century, the *multiband* extension of the BCS theory with dominant interband coupling remains invalid at arbitrarily small nonzero coupling strengths [31]. This inadequacy of the multiband weak-coupling theory was traced back to the neglect of the renormalization in the BCS theory (it assumes that $Z(\omega) = 1$), which imparts a completely different functional dependence on the coupling strength to the order parameter in the case of interband pairing.

In respect of the electromagnetic properties of superconductors, Mattis and Bardeen derived an expression for the optical conductivity already in 1958 [32]. However, it was obtained in the dirty limit $\gamma \gg 2\Delta$, where γ is the impurity scattering rate. Since for most conventional superconductors the superconducting energy gap Δ is very small, the theory was quite successful in reproducing the experimentally obtained optical response. Needless to say, for a very clean or a very strongly coupled superconductor this approximation becomes invalid. In the wake of the discovery of high-temperature superconductivity in the cuprates, this Mattis-Bardeen theory was extended by Zimmermann to arbitrary impurity scattering rates while remaining a weak-coupling theory [32] but even this extended theory is inapplicable to very pure strongly coupled superconductors, whose optical response must then be calculated in the framework of the complete Eliashberg theory as described in Refs. [28, 29].

As it so happens, the highest-quality (and the first available in this quality) superconducting iron-pnictide $Ba_{0.6}K_{0.4}Fe_2As_2$ was found to be precisely in this least convenient clean and strong-coupling limit [33], rendering the simpler Mattis-Bardeen and Zimmermann results inapplicable. The extreme purity of this compound, at first sight having only negative effect on the description of its optical properties by making most approximations inadequate, in fact reveals an optical signature of the boson mediating the superconducting pairing [22], which remains disguised in the dirty and/or weak-coupling limit, thus giving spectroscopic measurements much higher predictive power. This topic will be considered in more detail in Sect. 4.2.

References

1. Mahan, G. D. (2000). *Many particle physics*. Berlin: Springer.
2. Tompkins, H. G., & Irene, E. A. (2005). *Handbook of ellipsometry*. Norwich: William Andrew Inc.
3. Wagner, W., Riethmann, T., Feistel, R., & Harvey, A. H. (2011). New equations for the sublimation pressure and melting pressure of H_2O ice Ih. *Journal of Physical and Chemical Reference Data*, *40*, 043103.
4. Feistel, R., & Wagner, W. (2007). Sublimation pressure and sublimation enthalpy of H_2O ice Ih between 0 and 273.16 K. *Geochimica et Cosmochimica Acta*, *71*, 36–45.
5. Jackson, J. D. (1998). *Classical electrodynamics*. New York: Wiley.
6. Jakopic, G., & Papousek, W. (2000). Unified analytical inversion of reflectometric and ellipsometric data of absorbing media. *Applied Optics*, *39*, 2727.
7. Kawata, S., & Inouye, Y. (1995). Scanning probe optical microscopy using a metallic probe tip. *Ultramicroscopy*, *57*, 313–317.
8. Keilmann, F., & Hillenbrand, R. (2008). *Nano-optics and near-field optical microscopy*. London: Artech House.
9. Huth, F., Schnell, M., Wittborn, J., Ocelic, N., & Hillenbrand, R. (2011). Infrared-spectroscopic nanoimaging with a thermal source. *Nature Mater*, *10*, 352–356.
10. Cvitkovic, A., Ocelic, N. & Hillenbrand, R. (2007). Analytical model for quantitative prediction of material contrasts in scattering-type near-field optical microscopy. *Optics Express*, *15*, 8550–8565.
11. Sonier, J. E. (2001). μSR brochure. Retrieved from http://muon.neutron-eu.net/muon/files/musrbrochure.pdf
12. Yaouanc, A., & de Réotier, P. D. (2011). *Muon spin rotation, relaxation, and resonance: Applications to condensed matter*. Oxford: Oxford University Press.
13. Tinkham, M. (1995). *Introduction to superconductivity*. New York: McGraw-Hill.
14. Sonier, J. E., Brewer, J. H., & Kiefl, R. F. (2000). μSR studies of the vortex state in type-II superconductors. *Reviews of Modern Physics*, *72*, 769–811.
15. Niedermayer, Ch., Forgan, E. M., Glückler, H., Hofer, A., Morenzoni, E., Pleines, M., et al. (1999). Direct observation of a flux line lattice field distribution across an $YBa_2Cu_3O_{7-\delta}$ surface by low energy muons. *Physical Review Letters*, *83*, 3932.
16. Ashcroft, N. W., & Mermin, N. D. (1976). *Solid state physics*. Fort Worth: Cengage Learning EMEA.
17. Drude, P. (1900). Zur Elektronentheorie der Metalle. *Annals of Physics*, *306*, 566–613.
18. Drude, P. (1900). Zur Elektronentheorie der Metalle; II. Teil. Galvanomagnetische und thermomagnetische Effecte. *Annals of Physics*, *308*, 369–402.

19. Glover, R. E., & Tinkham, M. (1956). Transmission of superconducting films at millimeter-microwave and far infrared frequencies. *Physical Review*, *104*, 844–845.
20. Hirsch, J. (1992). Apparent violation of the conductivity sum rule in certain superconductors. *Physica C*, *199*, 305–310.
21. Hirsch, J. E., & Marsiglio, F. (2000). Optical sum rule violation, superfluid weight, and condensation energy in the cuprates. *Physical Review B*, *62*, 15131–15150.
22. Charnukha, A., Dolgov, O. V., Golubov, A. A., Matiks, Y., Sun, D. L., Lin, C. T., et al. (2011). Eliashberg approach to infrared anomalies induced by the superconducting state of $Ba_{0.68}K_{0.32}Fe_2As_2$ single crystals. *Physical Review B*, *84*, 174511.
23. Gor'kov, L. P. (1958). On the energy spectrum of superconductors. *Soviet Physics—JETP*, *7*, 505.
24. Migdal, A. B. (1958). Interaction between electrons and lattice vibrations in a normal metal. *Soviet Physics—JETP*, *7*, 996.
25. Eliashberg, G. (1960). Interactions between electrons and lattice vibrations in a superconductor. *Soviet Physics—JETP*, *11*, 696.
26. Grimvall, G. (1981). *The electron-phonon interaction in metals*. New York: E. P. Wohlfarth.
27. Carbotte, J. P. (1990). Properties of boson-exchange superconductors. *Reviews of Modern Physics*, *62*, 1027–1157.
28. Nam, S. B. (1967). Theory of electromagnetic properties of superconducting and normal systems. I. *Physical Review*, *156*, 470.
29. Nam, S. B. (1967). Theory of electromagnetic properties of strong-coupling and impure superconductors. II. *Physics Review*, *156*, 487.
30. Suhl, H., Matthias, B. T., & Walker, L. R. (1959). Bardeen-Cooper-Schrieffer theory of superconductivity in the case of overlapping bands. *Physical Review Letters*, *3*, 552–554.
31. Dolgov, O. V., Mazin, I. I., Parker, D., & Golubov, A. A. (2009). Interband superconductivity: Contrasts between Bardeen-Cooper-Schrieffer and Eliashberg theories. *Physical Review B*, *79*, 060502.
32. Mattis, D. C., & Bardeen, J. (1958). Theory of the anomalous skin effect in normal and superconducting metals. *Physical Review*, *111*, 412–417.
33. Popovich, P., Boris, A. V., Dolgov, O. V., Golubov, A. A., Sun, D. L., Lin, C. T., et al. (2010). Specific heat measurements of $Ba_{0.68}K_{0.32}Fe_2As_2$ single crystals: evidence for a multiband strong-coupling superconducting state. *Physical Review Letters*, *105*, 027003.

Chapter 4
Results and Discussion

It doesn't matter how beautiful your theory is, it doesn't matter how smart you are. If it doesn't agree with experiment, it's wrong.

—Richard Feynman

4.1 Criticality-Induced Optical Anomalies in 122 Iron Arsenides

4.1.1 Introduction

The standard Bardeen-Cooper-Schrieffer (BCS) theory of superconductivity, based solely on an effective attractive interaction between electrons mediated by phonons, does not provide a satisfactory explanation of the properties of strongly correlated high-temperature superconductors.[1] Theoretical proposals going back many years suggest that electronic excitations might enhance this interaction and thus contribute to the formation of the superconducting condensate [2–6]. These proposals appeared to gain some ground with the observation of superconductivity-induced transfer of the optical spectral weight in the cuprate high-temperature superconductors which involves a high-energy scale extending to the visible range of the spectrum [7]. In spite of numerous studies (for a comprehensive list of references see Ref. [8]) no modification of interband optical transitions in the superconducting state has been directly identified in the cuprates. Instead, the observed superconductivity-induced anomalies in the optical response of highly conducting CuO_2 planes were found to be confined to the energy range corresponding to transitions within the conduction band below the plasma edge. These changes are dominated by the narrowing of the broad Drude peak caused by a superconductivity-induced modification of the scattering

[1] The discussion in this section is largely based on Ref. [1].

A. Charnukha, *Charge Dynamics in 122 Iron-Based Superconductors*, Springer Theses, DOI: 10.1007/978-3-319-01192-9_4,

rate [9–11]. A minute redistribution of the spectral weight between the conduction band and high-energy Hubbard bands generated by Coulomb correlations may also play a role [12, 13].

Current research on the recently discovered iron-pnictide superconductors [14] suggests that electronic correlations are weaker than those in the cuprates. Unlike in the cuprates, the Fermi surface has been reliably determined over the entire phase diagram and shows good agreement with density functional calculations (see Sect. 2.3). The superconducting state of the iron pnictides appears to fit well into a BCS framework in which phonons, which in these compounds interact only weakly with electrons [15], are replaced by spin fluctuations [16]. The ellipsometric data we present here are consistent with the hypothesis that electronic correlations only result in a modest renormalization of the electronic states. However, the superconductivity-induced optical anomalies we observed involve modification of an absorption band peaked at an energy of 2.5 eV, two orders of magnitude larger than the superconducting gap $2\Delta \approx 20$ meV. In contrast to the cuprate superconductors, this high-energy anomaly has a regular Lorentzian shape in both the real and imaginary part of the dielectric function and is confined to energies well above the plasma edge $\hbar\omega_{\mathrm{pl}} \approx 1.5$ eV. It can be explained as a consequence of non-conservation of the total number of unoccupied states involved in the corresponding optical transitions due to the opening of the superconducting gap. This implies that unconventional interactions beyond the BCS framework must be considered in models of the superconducting pairing mechanism.

4.1.2 Superconductivity-Induced Optical Anomalies in $Ba_{0.68}K_{0.32}Fe_2As_2$

The measurements were carried out on a single crystal of $Ba_{1-x}K_xFe_2As_2$ (BKFA) with $x = 0.32$ and superconducting $T_c = 38.5$ K. Specific-heat measurements on the same sample confirm its high purity and the absence of secondary electronic phases [17]. We performed direct ellipsometric measurements of the in-plane complex dielectric function $\varepsilon(\omega) = \varepsilon_1(\omega) + i\varepsilon_2(\omega) = 1 + 4\pi i\sigma(\omega)/\omega$ over a range of photon energies extending from the far infrared ($\hbar\omega = 12$ meV) to the ultraviolet ($\hbar\omega = 6.5$ eV) with a subsequent KK consistency analysis (see also Sect. 3.5). The far-infrared optical conductivity is dominated by the opening of a superconducting gap of $2\Delta \approx 20$ meV below T_c (Fig. 4.1a), in accordance with the previous studies of the optimally doped BKFA [18]. The low-energy missing area in the optical conductivity spectrum below T_c, $\delta A_{\mathrm{L}} = \int_{0+}^{10\Delta}(\sigma_1^{40\,\mathrm{K}}(\omega) - \sigma_1^{10\,\mathrm{K}}(\omega))d\omega$, is contained within 10Δ and amounts to $\omega_{\mathrm{pl}}^{\mathrm{sc}} = \sqrt{8\delta A_{\mathrm{L}}} = 0.9$ eV, equivalent to a London penetration depth of $\lambda_{\mathrm{p}} = 2200$ Å. The fraction of the missing area below 12 meV not accessible to the experiment was accurately quantified from the requirement of KK consistency of the independently measured real and imaginary part of the dielectric function (as discussed in Sect. 3.5).

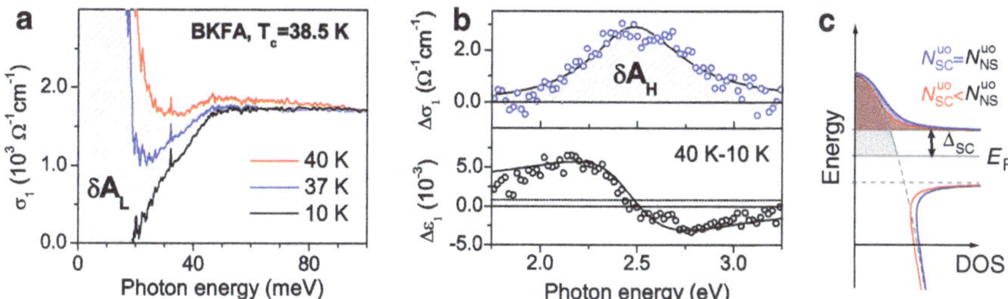

Fig. 4.1 **a** Real part of the far-infrared optical conductivity of $Ba_{0.68}K_{0.32}Fe_2As_2$ and the missing area. **b** Difference spectra of the real part of the optical conductivity (*top panel*) and dielectric function (*bottom panel*) between 40 and 10 K, with a small background shift (*horizontal dashed line*) detected by temperature modulation measurements shown in Fig. 4.2. Lorentzian fit to both spectra (*black solid lines*). **c** Density of states in the normal (*gray dashed line*), conventional superconducting state (*blue line*), and an unconventional state with a depletion of unoccupied states (*red line*). *Filled areas* of respective colors represent the total number of unoccupied states. Figure adapted by permission from Macmillan Publishers Ltd: Nature Communications Ref. [1], copyright (2011)

Careful examination of the visible range uncovered a superconductivity-induced suppression of an absorption band at 2.5 eV. Figure 4.1b shows difference spectra between 40 and 10 K of the real part of the optical conductivity and dielectric function. The suppressed band has a Lorentzian lineshape and is modified abruptly across the superconducting transition, consistently in both $\Delta\sigma_1$ and $\Delta\varepsilon_1$, as shown in Fig. 4.2b for $\sigma_1(2.5\,\mathrm{eV})$. The temperature dependence of the suppression (blue open circles) coincides with that of the far-infrared conductivity due to the opening of the superconducting gap (red filled circles). Thus the onset of superconductivity not only modifies the low-energy quasiparticle response, but also affects the overall electronic structure including interband transitions in the visible range of the spectrum. Since the spectral-weight (SW) loss δA_H is not balanced in the vicinity of the absorption band (Fig. 4.1b), our data indicate a SW transfer over a wide energy range. We note that the superconductivity-induced modification of the lattice parameters only results in a minute volume change of $\Delta V/V \approx 5 \times 10^{-7}$, which is insufficient to explain the optical anomaly (C. Meingast, personal communication). A KK consistency analysis could not be carried out with sufficient accuracy to show whether or not the SW liberated from the absorption band at the superconducting transition contributes to the response of the superconducting condensate at zero energy. We did, however, detect a minute rise of the background level of $\varepsilon_1(1.5\text{–}3.5\,\mathrm{eV})$ below T_c, which according to the KK relation implies that the SW is transferred to energies below 1.5 eV. This effect was identified from a simultaneous fit of $\Delta\sigma_1(\omega)$ and $\Delta\varepsilon_1(\omega)$ (horizontal dashed line in the bottom panel of Fig. 4.1b). To estimate the background shift more accurately, temperature modulation measurements of σ_1 at the resonance photon energy 2.47 eV and ε_1 at off-resonance photon energies of 2.12 and 2.82 eV were carried out, as shown in Fig. 4.2. The sample temperature was changed between 20 and 40 K with a period of 1800 s and later averaged over 24

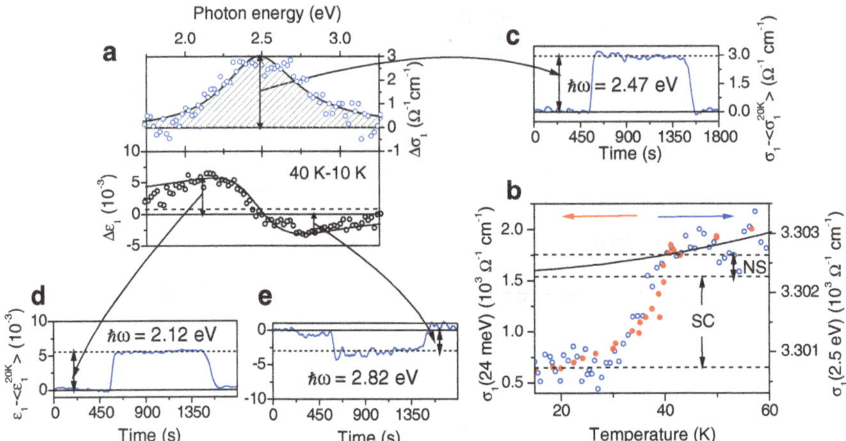

Fig. 4.2 **a** Difference spectrum of the real part of the optical conductivity (*top panel*) and the dielectric function (*bottom panel*) between 40 and 10 K (*blue* and *black open circles*, respectively). Lorentzian fit to both spectra (*black solid lines*). Small background shift (*dashed line* in the *bottom panel*) obtained from the fit and further confirmed by temperature modulation measurements **c**–**e**. **b** Temperature scan of the real part of the optical conductivity between 60 and 15 K at 2.5 eV (*blue open circles*, right scale) and 24 meV (*red filled circles*, left scale). The contribution of the normal-state dynamics (*black solid line*) was estimated to determine the magnitude of the SC-induced jump (*dashed lines*). **c**–**e** Temperature modulation of the real part of the optical conductivity (**c**) and dielectric function (**d**, **e**) between 20 and 40 K with a period of 1800 s averaged over 24 waveforms. Figure adapted by permission from Macmillan Publishers Ltd: Nature Communications Ref. [1] and supplementary material therein, copyright (2011)

periods to reduce noise to $\Delta\varepsilon_1 = 10^{-4}$. This confirms a minute background increase of $\Delta\varepsilon_1(2.47\,\text{eV}) = (8 \pm 4)\ 10^{-4}$.

Since the spectral weight δA_{H} liberated from the absorption band upon cooling below T_{c} comprises only $\sim$0.5 % of the total SW δA_{L} of the superconducting condensate, its contribution to the low-energy charge dynamics might be considered negligible. However, assuming that this additional high-energy SW contributes to the itinerant-carrier response below T_{c}, a simple estimate in the framework of the tight-binding nearest-neighbor approximation [19, 20] shows that this would lead to a reduction of the electronic kinetic energy of 0.60 meV/unit cell in the superconducting state. To demonstrate this, we assume that the missing SW is fully transferred to the intraband itinerant response. Then, following Refs. [19, 20], one obtains: $\Delta SW(\Omega) = (\pi e^2 a^2/2\hbar^2 V_u)\langle -E_{\text{K}}\rangle$, where $SW(\Omega) = \int_{0^+}^{\Omega} \sigma_1(\omega)d\omega$ is the in-plane SW, a is the in-plane lattice constant, V_u—unit cell volume and $\langle -E_{\text{K}}\rangle = (1/N)\sum_{\mathbf{k},\sigma} n_{\mathbf{k},\sigma}\partial^2\varepsilon_{\mathbf{k}}/\partial\mathbf{k}^2$ (where N is the total number of electrons, $n_{\mathbf{k},\sigma}$ is the density of electrons with wave vector $\vec{k}$ and spin σ, and $\varepsilon_{\mathbf{k}}$ is the electronic band dispersion) is a measure of the system's kinetic energy per unit cell. In the case of the tight-binding nearest-neighbor approximation it is exactly equal to the kinetic energy of the charge carriers per unit cell and thus contributes to the condensation energy directly. This approximation might not hold for the iron-pnictide superconductors but the above estimate shows that though small the additional high-energy SW in the superconducting $Ba_{0.68}K_{0.32}Fe_2As_2$ is enough to account for the condensation energy in this compound $\Delta F(0) = 0.36$ meV/unit cell obtained from specific-heat measurements on the same sample [17]. It is thus important to establish the origin of this unusual optical anomaly.

4.1.3 LDA Calculations

We therefore compared our data to the results of ab-initio electronic-structure calculations in the framework of LDA (Figs. 4.3a, b), performed using the linear-muffin-tin orbital method in the atomic sphere approximation [21] starting form the known

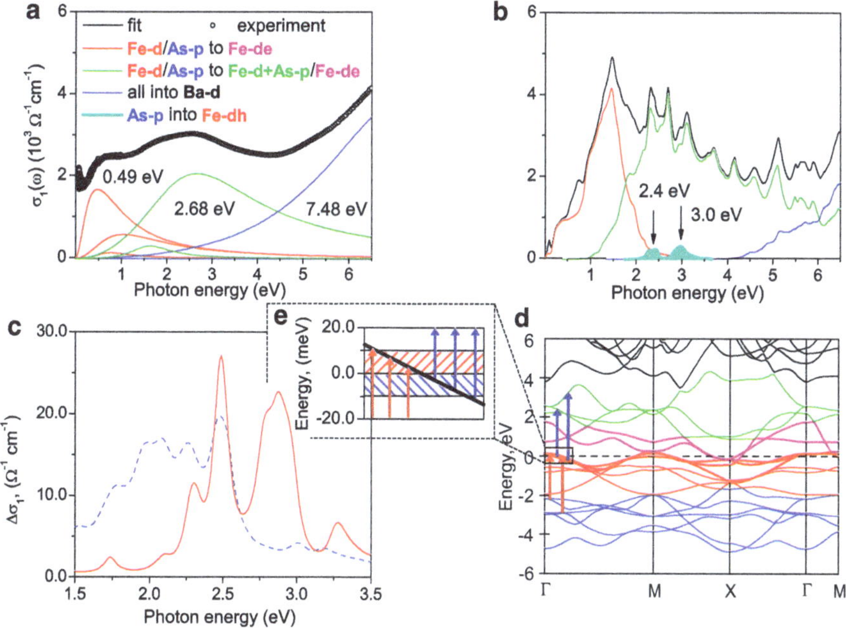

Fig. 4.3 **a** Real part of the optical conductivity of $Ba_{0.68}K_{0.32}Fe_2As_2$ and contributing interband transitions determined by a dispersion analysis. **b** Corresponding LDA calculation with a breakdown into separate orbital contributions described in the legend **a**. **c** Difference real part of the optical conductivity between ungapped and gapped regimes for transitions to (*red solid line*) and from (*blue dashed line*) the Fermi level. **d** Band structure from the same LDA calculation as in **b**. Color coding of the dispersion curves corresponds to the *text* color in legend **a**. The optical transitions at the Γ point of the Brillouin zone with a contribution to the optical conductivity shown in **c** are represented schematically as *red* and *blue arrows*. **e** Population (*red*) or depletion (*blue*) of all electronic states within one superconducting gap $\Delta = 10$ meV. *Arrows* depict the same optical transitions as in **d**. Figure adapted by permission from Macmillan Publishers Ltd: Nature Communications Ref. [1] and supplementary material therein, copyright (2011)

crystal structure of $Ba(Sr)_{1-x}K_xFe_2As_2$ [22, 23]. A dispersion analysis of the experimental optical conductivity in the range 0.5–6.5 eV yielded three major interband transitions in $Ba_{0.68}K_{0.32}Fe_2As_2$. Comparison to the LDA results enabled us to identify the initial and final states of these transitions. The lowest-energy transition is located at about 1 eV (red line) and stems from intraband Fe-d and interband As-p to Fe-d transitions. The major contribution to the optical response in the visible spectral range comes from transitions starting from Fe-d or As-p orbitals into strongly hybridized Fe-d, As-p or Fe-d orbitals (green line). Finally, the UV absorption comes from higher-energy transitions into Ba-d states (blue line).

Although the high-energy electronic structure of BKFA is predicted quite well by the LDA calculations, the experimental quasiparticle response due to transitions within the conduction band (or, given the multiorbital structure of the iron pnictides, a narrowly spaced set of conduction bands) shows a significant deviation. The discrepancy can be quantified by the squared ratio of the band-structure plasma frequency $\omega_{\mathrm{pl}}^{\mathrm{LDA}} = 2.7\,\mathrm{eV}$ (not included in Fig. 4.3b) to its experimental counterpart $\omega_{\mathrm{pl}}^{\mathrm{exp}} = 1.5\,\mathrm{eV}$, which can be obtained in practice from the residual optical response, after the interband transitions identified using the dispersion analysis have been subtracted. This ratio approximates the quasiparticle effective-mass renormalization factor $m^{\star}/m_{\mathrm{band}} = \left(\omega_{\mathrm{pl}}^{\mathrm{LDA}}/\omega_{\mathrm{pl}}\right)^2 \approx 3$. Such an enhancement is consistent with de Haas-van Alphen and ARPES experiments on other compounds of the 122 family [24–26] and was recently reproduced by combined LDA+DMFT calculations for both 1111 and 122 compounds [27]. These calculations do not show evidence for the formation of Hubbard bands and thus indicate moderate electron-electron correlations. This explains the good agreement of the LDA optical conductivity above 1.5 eV with the experimental data.

Now we turn to the physical origin of the superconductivity-suppressed absorption band. The same LDA calculation revealed a set of interband transitions centered at 2.5 eV, which originate or terminate in states exhibiting hole dispersion and crossing the Fermi level at the Γ and M points of the Brillouin zone (Fig. 4.3d). We confidently assign these states to Fe-$d_{yz,zx}$ and Fe-d_{xy} orbitals. The other states involved in these transitions belong to As-$p_{x,y}$/Fe-d_{z^2} hybrid orbitals about 2–3 eV below and above the Fermi level, giving rise to a bandwidth of $\Delta E \approx 1\,\mathrm{eV}$, in remarkable agreement with experiment. Suppression of the absorption band over its full width can be explained by a redistribution of the occupation of Fe-$d_{yz,zx}$ and Fe-d_{xy} states under the Fermi level below the superconducting transition. This mechanism is supported by LDA calculations in which the DOS within one superconducting gap energy above or below the Fermi level was eliminated, leading to the observed suppression of the optical transitions shown in cyan in Fig. 4.3b. This result is demonstrated in Fig. 4.3c–e. To substantiate that a redistribution of the occupied states below the Fermi level within one superconducting gap can account for the observed suppression, the optical conductivity in the LDA framework was calculated for the two cases of hole and electron transitions in the vicinity of Γ and M points of the Brillouin zone, i.e. for the hole pockets of the Fermi surface, shown in Fig. 4.3c, d. Shown in Fig. 4.3e are the difference optical conductivity spectra between gapped and ungapped regime. In the gapped case the transitions in the energy window $\Delta E = \Delta = 10\,\text{meV}$ below (blue line) and above (red line) the Fermi level were forbidden, to simulate the complete depletion (population) of the occupied (empty) states. Hole (electron) contribution to the optical conductivity displays a suppression only when the states above (below) the Fermi level are fully populated (depopulated) to within one superconducting gap. The size of the effect in both cases is an order of magnitude larger than the experimentally detected suppression. Therefore a fractional population redistribution can indeed account for the superconductivity-induced anomaly in $Ba_{0.68}K_{0.32}Fe_2As_2$.

4.1.4 Spin-Density-Wave–Induced Spectral Weight Transfer in $SrFe_2As_2$

We have further explored the validity of this scenario by repeating our ellipsometric measurements on several parent compounds of the 122 family of the iron-arsenide superconductors. Since the SDW instability exhibited by these compounds is also believed to be induced by the nesting of electronic states on different electronic bands, we expect a similar optical anomaly at the SDW transition as the one we observed in the superconductor. The magnitude of the anomaly is expected to be larger than the one in the superconductor, because the SDW transition occurs at a higher temperature and generates a larger energy gap. In $SrFe_2As_2$ (SFA), we indeed find a strong reduction of optical absorption upon cooling below $T_{\text{SDW}} = 200\,\text{K}$. The difference spectra of $\varepsilon_1(\omega)$ and $\varepsilon_2(\omega)$ between 200 and 175 K show a double-peak structure with maxima at 2.4 and 3.4 eV (Fig. 4.4a). A temperature scan across T_{SDW} at the frequency of the second peak (inset of Fig. 4.4a) further confirms that this effect is induced by the formation of SDW. A direct comparison with the superconducting compound is complicated by a pronounced modification of the electronic structure due to the coincident magnetic and structural transitions. Nevertheless, certain information can be gained from the critical behavior of the in-plane spectral weight $\Delta SW(\Omega) = \int_0^{\Omega} \Delta\sigma_1(\omega)d\omega$. Figure 4.4b shows difference SW of $SrFe_2As_2$ between 200 K and 175 K as a function of the upper integration limit Ω. The change of the SW in the extrapolation region below 12 meV was accurately determined via a KK consistency analysis (see Sect. 3.5), as illustrated in the inset of Fig. 4.4b. With $\Delta SW(12\,\text{meV}) = 0$ across the transition (blue filled circle) the KK transformation of $\Delta\varepsilon_2(\omega)$ (blue line) deviates significantly from the experimentally measured $\Delta\varepsilon_1(\omega)$. Gradually increasing this SW brings them closer together until they finally coincide (red line) thus fixing $\Delta SW(12\,\text{meV}) = 0.015\,\text{eV}^2$ (red filled circle). The higher-energy redistribution of the SW is broken down in Fig. 4.4b into regions of SW gain (blue areas) and loss (red areas) in the SDW compared to the paramagnetic state. The SW lost due to the opening of the SDW gap [28] (the first red region) is partly transferred to the electronic excitations across the gap (the first blue region) and fully recovered by 1.5 eV. These processes are then followed by higher-energy redistribution in the region from 1.5 to 4.0 eV involving the SW of

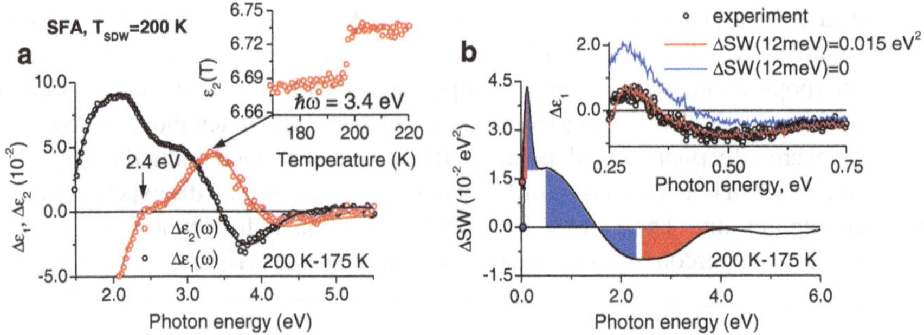

Fig. 4.4 **a** Difference spectra of the real and imaginary part of the dielectric function of $SrFe_2As_2$ between 200 K and 175 K. Lorentzian fit to both (*solid lines*). (Inset) Temperature scan of ε_2 at 3.4 eV. **b** SW redistribution between 200 and 175 K. SW in the extrapolation region below 12 meV before (*blue filled circle*) and after (*red filled circle*) a KK consistency check. *Blue* and *red filled areas* represent regions of SW gain and loss, respectively, in the antiferromagnetic versus the normal state. (Inset) Difference spectra of the real part of the dielectric function obtained experimentally (*open circles*) and from the KK transformation of the real part of the optical conductivity (*solid lines*, colors match *filled circles*). Figure adapted by permission from Macmillan Publishers Ltd: Nature Communications Ref. [1], copyright (2011)

the suppressed bands. It appears unlikely that such a high-energy SW transfer could result from a modification of the electronic structure due to the magnetic transition because the effects of the electronic reconstruction at the SDW transition are limited by 1.5 eV. A modification of the matrix elements at the structural transitions of sufficient strength cannot account for the observed suppression, because this would be accompanied by an even larger effect at higher energies, clearly absent in Fig. 4.4b. A redistribution of charge carriers between the SDW-coupled bands analogous to that in the superconducting compound provides a more natural explanation. The same physical reasons might explain the orbital polarization that breaks the degeneracy of Fe-d_{xz} and Fe-d_{yz} orbitals recently observed in the Ba-based parent of the same family by photoemission spectroscopy [29].

4.1.5 Discussion

The population redistribution required to account for the superconductivity-induced reduction of absorption in BKFA is, however, at variance with the conventional theory of superconductivity. In the framework of the standard BCS approach, the opening of an energy gap in a single-band superconductor leads to a bending of the quasiparticle dispersion and an expulsion of the DOS in the vicinity of the Fermi surface (gray and blue areas in Fig. 4.1c) [30], with the total number of unoccupied states below the transition conserved $N^{uo}_{SC} = N^{uo}_{NS}$ (blue area is equal to the gray area in Fig. 4.1c). This can only lead to a small corrugation of an optical absorption band on the scale of one superconducting-gap energy superimposed on the overall broad feature without any modification of its spectral weight [31]. The experimentally observed suppression of an absorption band *on the scale of its full width* necessarily requires a population imbalance $N_{SC}^{uo} < N_{NS}^{uo}$ (red area unequal to the gray area in Fig. 4.1c). This effect can be clearly identified as a consequence of superconductivity because the temperature dependence of the suppression mimics that of the optical conductivity in the FIR region due to the opening of the superconducting gap, as shown in Fig. 4.2b.

All of the iron-pnictide superconductors are known to have multiple superconducting gaps [14] and theoretical work indicates a dominant contribution of electron pairing between different bands to the formation of the superconducting state [16]. Redistribution of the occupation of different bands below T_c could explain the optical anomaly we observed (see Fig. 4.5). However, even a generalization of the standard BCS theory to the multiband case [32] does not take this effect into account. It requires a lowering of the material's chemical potential in the superconducting state, as discussed below. In the presence of large Fe–As bond polarizability [33, 34] it can potentially enhance superconductivity in the iron pnictides.

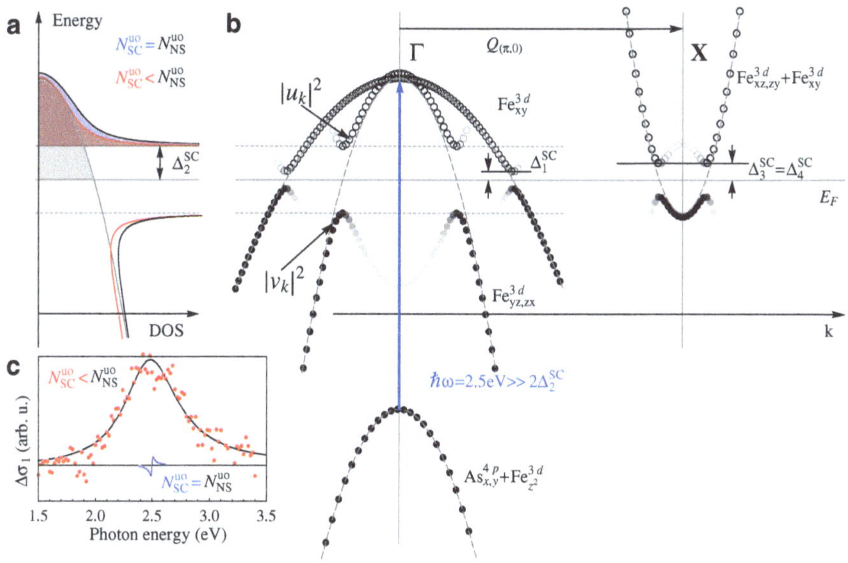

Fig. 4.5 **a** Density of states in the normal (*gray line*), conventional superconducting state (*black line*), and an unconventional state with a depletion of unoccupied states (*red line*). *Filled areas* of respective colors represent total number of unoccupied states. **b** Schematic representation of the band structure of BKFA in the normal (*dashed parabolas*) and the superconducting state (*filled* and *empty circles* for the occupied with probability $|v_{\mathbf{k}}|^2$ and the unoccupied with probability $|u_{\mathbf{k}}|^2$ states with a Bogolubov dispersion, respectively). The pair $(u_{\mathbf{k}}, v_{\mathbf{k}})$ is unique for each separate band in the conventional multiband BCS approach [32, 35]. The orbitals of dominant contribution to each particular band are specified. **c** (Schematic) Difference spectra of the real part of the optical conductivity between 40 and 10 K. Figure adapted by permission from Macmillan Publishers Ltd: Nature Communications Ref. [1] (supplementary material), copyright (2011)

Interactions of electrons in different energy bands at the Fermi level may provide a common framework for an explanation of the optical anomalies in the spin-density-wave and superconducting compounds. It is important to note that these anomalies affect only a small fraction of the interband transitions, which involve initial states of As-p-orbital character deep below the Fermi level. This indicates that these orbitals significantly influence electronic instabilities in the iron arsenides, possibly due to the high polarizability of the As–Fe bonds. Our study points to an optical SW transfer from high energies to below 1.5 eV induced by collective electronic instabilities. In the superconductor, it occurs at energies two orders of magnitude larger than the superconducting gap energy, suggesting that electronic pairing mechanisms contribute to the formation of the superconducting condensate.

Finally, we would like to make a few remarks regarding the interband optical conductivity in the single- and multiband BCS theory. In the framework of the BCS theory the charge carriers and elementary excitations in the superconducting state differ significantly from those of in the normal state. This is manifested in the modified with respect to free charge carriers dispersion of the excitations of the superconducting condensate—Bogolubov quasiparticles. To obtain this dispersion one can introduce quasiparticle operators diagonalizing the original BCS Hamiltonian [30, 35]

$$\mathscr{H} = \sum_{\mathbf{k}\sigma} \epsilon_{\mathbf{k}} n_{\mathbf{k}\sigma} + \sum_{\mathbf{kl}} V_{\mathbf{kl}} c^{\dagger}_{\mathbf{k}\uparrow} c^{\dagger}_{-\mathbf{k}\downarrow} c_{-\mathbf{l}\downarrow} c_{\mathbf{l}\uparrow}$$

as follows:

$$\begin{aligned} c_{\mathbf{k}\uparrow} &= u^*_{\mathbf{k}} \gamma_{\mathbf{k}0} + v_{\mathbf{k}} \gamma^{\dagger}_{\mathbf{k}1}, \\ c_{-\mathbf{k}\downarrow} &= -v^*_{\mathbf{k}} \gamma_{\mathbf{k}0} + u_{\mathbf{k}} \gamma^{\dagger}_{\mathbf{k}1}. \end{aligned} \tag{4.1}$$

The complex functions of the k-vector $u_{\mathbf{k}}$ and $v_{\mathbf{k}}$ determine the probability of the pair state comprised of electrons with momenta **k** and $-\mathbf{k}$ being empty or occupied, respectively:

$$\begin{aligned} \left|u_{\mathbf{k}}\right|^2 &= \frac{1}{2}\left(1 + \frac{\xi_{\mathbf{k}}}{E_{\mathbf{k}}}\right), \\ \left|v_{\mathbf{k}}\right|^2 &= \frac{1}{2}\left(1 - \frac{\xi_{\mathbf{k}}}{E_{\mathbf{k}}}\right), \end{aligned} \tag{4.2}$$

where $E_{\mathbf{k}} = \sqrt{\xi^2_{\mathbf{k}} + \Delta^2}$ is the Bogolubov quasiparticle's dispersion and $\xi_{\mathbf{k}} = \epsilon_{\mathbf{k}} - \mu$ is the normal-state electron dispersion measured with respect to the chemical potential μ. This probability distributions are inherently smeared around the Fermi level in the ground state. This smearing of the quasiparticle occupation probabilities at 0 K closely resembles that of normal-state particles at $T = T_{\mathrm{c}}$ [30].

One of the important and quite intuitive consequences of the Bogolubov-Valatin transformation (4.1) is that the operators $\gamma_{\mathbf{k}\sigma}$ and $c_{\mathbf{k}\sigma}$ are connected via a unique, one-to-one relation. This immediately implies the conservation of the total number of states in a given energy range, i.e.

$$N_{\text{SC}}(E)dE = N_{\text{NS}}(\xi)d\xi, \tag{4.3}$$

where $N_{\text{SC}}(E)$ is the quasiparticle DOS in the superconducting state, $N_{\text{NS}}(\xi)$ is the normal-state electron DOS. This relation requires that the states within one Δ below the Fermi level be expelled to energies lower than $E_{\text{F}} - \Delta$, while those within one Δ above the Fermi level to energies higher than $E_{\text{F}} + \Delta$, as illustrated schematically in Fig. 4.5a. This process conserves the population above and below the Fermi level so that $N_{\text{SC}}^{\text{uo}} = N_{\text{NS}}^{\text{uo}}$ (blue area equal to the gray area in Fig. 4.5a). The exact analytical expression for the DOS of Bogolubov quasiparticles follows from equation (4.3) bearing in mind the definition of the quasiparticle dispersion $\xi_{\mathbf{k}} = \sqrt{E_{\mathbf{k}}^2 - \Delta^2}$:

$$N_{\text{SC}}(E) = N_{\text{NS}}(\xi)\frac{d\xi}{dE} = N_{\text{NS}}(\xi(E))\frac{E_{\mathbf{k}}}{\sqrt{E_{\mathbf{k}}^2 - \Delta^2}}. \tag{4.4}$$

This expression is plotted in Fig. 4.5a (black solid line) for the case of a free-electron normal-state dispersion (gray dashed line). The square-root singularity in the quasiparticle DOS stems from the flattening of the normal-state dispersion in the vicinity of the Fermi surface, as shown in Fig. 4.5b (the inner hole dispersion corresponds to the DOS plotted in Fig. 4.5a). Occupied quasiparticle states are depicted as filled black circles, while the quasiparticle vacancies are shown as empty black circles. The fading of filled and empty black circles represents the occupation probabilities ($|u_{\mathbf{k}}|^2$, $|v_{\mathbf{k}}|^2$) in equation (4.2). As it has already been mentioned above there exists a finite smearing of these probabilities even at 0 K. It leads to finite occupation of those regions of the Brillouin zone unoccupied in the normal state, the so-called *backfolding* of the quasiparticle dispersion, clearly visible in Fig. 4.5b. As the optical conductivity involves averaging over reciprocal space it cannot resolve the backfolding as opposed to ARPES, where this effect has been reliably established [36–38]. On the other hand the effect of smearing itself is incorporated in the DOS within the BCS formalism and is, therefore, included into our considerations. It is a single-band effect of superconductivity on the band structure and does not lead to population redistribution, i.e. the total number of unoccupied states below the transition is conserved $N_{\text{SC}}^{\text{uo}} = N_{\text{NS}}^{\text{uo}}$ (blue area is equal to the gray area in Fig. 4.5a), and thus cannot explain the experimentally observed suppression of the high-energy SW.

To account for the multiband character of the iron pnictides one may consider the multiband BCS theory [32]. It is a straightforward generalization of the single-band BCS theory with the only complication that each separate band has its own superconducting energy gap, quasiparticle dispersion, and a pair ($u_{\mathbf{k}}$, $v_{\mathbf{k}}$). However, in the framework of this multiband theory the quasiparticle operators only mix the normal-state creation and annihilation operators of particles from the same band (preferring our notation to that of Ref. [32]):

$$\begin{aligned} c_{\mathbf{k}\uparrow} &= u_{\mathbf{k}}^{(c)*} e_{\mathbf{k}0} + v_{\mathbf{k}}^{(c)} e_{\mathbf{k}1}^{\dagger}, \\ c_{-\mathbf{k}\downarrow} &= -v_{\mathbf{k}}^{(c)*} e_{\mathbf{k}0} + u_{\mathbf{k}}^{(c)} e_{\mathbf{k}1}^{\dagger}, \\ d_{\mathbf{k}\uparrow} &= u_{\mathbf{k}}^{(d)*} f_{\mathbf{k}0} + v_{\mathbf{k}}^{(d)} f_{\mathbf{k}1}^{\dagger}, \\ d_{-\mathbf{k}\downarrow} &= -v_{\mathbf{k}}^{(d)*} f_{\mathbf{k}0} + u_{\mathbf{k}}^{(d)} f_{\mathbf{k}1}^{\dagger}, \end{aligned} \tag{4.5}$$

where $(c_{\mathbf{k}\uparrow}, d_{\mathbf{k}\uparrow})$ are normal-state particle operators and $(e_{\mathbf{k}\uparrow}, f_{\mathbf{k}\uparrow})$ are the multi-band counterparts of the operators $\gamma_{\mathbf{k}\uparrow}$ in the single-band BCS theory. Although coefficients $(u_{\mathbf{k}}^{(c,d)}, v_{\mathbf{k}}^{(c,d)})$ certainly depend on the properties of both bands via the interband coupling, the relations (4.5) are still unique one-to-one relations, which immediately implies that, however complicated the quasiparticle dispersions may be, for each separate band the relation

$$N_{\mathrm{SC}}^{(i)}(E)dE^{(i)} = N_{\mathrm{NS}}^{(i)}(\xi)d\xi, \tag{4.6}$$

holds, with (i) running through all bands. As a consequence, just like in the single-band case, the total occupied/unoccupied population is conserved across the superconducting transition and thus only changes of the interband optical conductivity on the scale of 2Δ are expected. The population imbalance $N_{\mathrm{SC}}^{\mathrm{uo}} < N_{\mathrm{NS}}^{\mathrm{uo}}$ (red area in Fig. 4.5a smaller than the gray area) required to accommodate the experimentally observed suppression of the 2.5 eV absorption band over its full width of about 1 eV can come from redistribution of occupation of different bands below T_{c}. It requires a lowering of the material's chemical potential in the superconducting state and, therefore, an additional contribution to the condensation energy. However, the standard BCS theory and its generalization to the multiband case do not self-consistently take into account this effect: though predicting a lowering of the chemical potential as a consequence of a non-zero gain in the free energy of the system (condensation energy), they premise on equations with an essentially constant chemical potential. Consistent treatment of a variable chemical potential might render the Bogolubov-Valatin transformation (4.5) inappropriate in the multiband case and violate the population conservation of the occupied and unoccupied states within each band (as shown in Fig. 4.5a)—a fundamental consequence of the standard BCS theory. The resulting correction, small as it may be for the conventional superconductors, in the presence of large Fe–As bond polarizability can lead to a large effect and potentially enhance superconductivity in the iron pnictides.

4.2 Eliashberg Description of Infrared Anomalies in $Ba_{0.68}K_{0.32}Fe_2As_2$

4.2.1 Introduction

The discovery of the iron-based superconductors[2] [40] has generated significant experimental and theoretical effort to unravel the mechanism of high-temperature superconductivity in these compounds. This effort has yielded a comprehensive experimental description of the electronic structure at the Fermi level, which includes multiple Fermi surface sheets in a good agreement with density functional calculations [41, 42]. Partial nesting between at least two of these sheets leads to a spin-density-wave instability that renders the metallic parent compounds antiferromagnetic. In the superconducting compounds spin fluctuations become the source of strong repulsive interband interactions and might give rise to superconductivity with different signs on these sheets [14].

The most incisive experimental data have been obtained on high-quality single crystals of the iron pnictides with the so-called 122 structure, for instance $BaFe_2As_2$, with K substituted for Ba or Co for Fe, resulting in hole and electron doping, respectively. In all of these materials five Fermi surface sheets have been identified in calculations and confirmed by numerous independent experimental studies [41, 42]: in the reduced Brillouin-zone scheme these are three hole pockets at the Γ point and two almost degenerate electron pockets at the M point with nesting between hole and electron sheets. Among all of these 122 materials, the optimally hole-doped compound $Ba_{0.68}K_{0.32}Fe_2As_2$ (BKFA) has the highest transition temperature of 38.5 K. Due to their exceptional quality, crystals of this compound are well suited as a testbed for theoretical models. A four-band Eliashberg theory with strong interband coupling has already proven successful in accounting for the transition temperature, as well as the temperature dependence of the free energy and superconducting gaps of this compound [17]. This analysis has made clear that a satisfactory description of the bulk thermodynamical properties in the superconducting state can only be obtained via the strong coupling of itinerant electrons to spin fluctuations or other bosons with the spectral weight below 50 meV.

In this work we extend the approach of Ref. [17] to describe the far-infrared properties of the same BKFA single crystal. We show that major characteristic features of superconductivity can be explained within a strong-coupling Eliashberg approach with two distinct values of the superconducting energy gap $2\Delta_A \approx 6\ k_B T_c$ and $2\Delta_B \approx 2.2\ k_B T_c$, in quantitative agreement with ARPES [36, 43–45], STS [46, 47] and specific-heat measurements [17]. We also demonstrate that within this approach the difference in the shape of the infrared conductivity spectra of hole- and electron-doped 122 compounds is reproduced by strong-coupling calculations in the clean and dirty limits (weak and strong impurity scattering), respectively.

[2] The discussion in this section is largely based on Ref. [39].

4.2.2 Experimental Details

The optimally-doped BKFA single crystals were grown in zirconia crucibles sealed in quartz ampoules under argon atmosphere [48]. From dc-resistivity, magnetization and specific-heat measurements we obtained $T_c = 38.5 \pm 0.2$ K. The sample surface was cleaved prior to every optical measurement. The full complex dielectric function $\varepsilon(\omega)$ was obtained in the range 0.01–6.5 eV using broadband ellipsometry, as described in Ref. [49]. In this work we focus on the itinerant charge carrier response contained within the far-infrared spectral range measured at the infrared beamline of the ANKA synchrotron light source at Karlsruhe Institute of Technology, Germany.

4.2.3 Far-Infrared Conductivity

The optical response of BKFA in the far-infrared spectral range is shown in Fig. 4.6a, b respectively for the real part of the optical conductivity $\sigma(\omega) = \sigma_1(\omega) + i\sigma_2(\omega)$ and dielectric function $\varepsilon(\omega) = 1 + 4\pi i\sigma(\omega)/\omega$. It is dominated by the contribution of itinerant charge carriers manifested in negative values of $\varepsilon_1(\omega)$. Figure 4.6a also reveals a peak around 15–20 meV in $\sigma_1(\omega)$ with a concomitant upturn in $\varepsilon_1(\omega)$ at higher temperatures indicating the presence of a collective excitation. The superconducting transition is accompanied by a suppression of the optical conductivity up to 50 meV ($14k_BT_c$). The corresponding missing area in $\sigma_1(\omega)$ between 41 and 10 K, $\int_{0+}^{6\Delta_A} \Delta\sigma_1(\omega)d\omega = (8\lambda_L^2)^{-1}$, manifests itself as a characteristic $-1/(\lambda_L\omega)^2$ contribution to $\varepsilon_1(\omega)$ in Fig. 4.6b in the superconducting state. The low-temperature London penetration depth $\lambda_L(10\text{ K}) = 2200$ Å extracted from these data is consistent

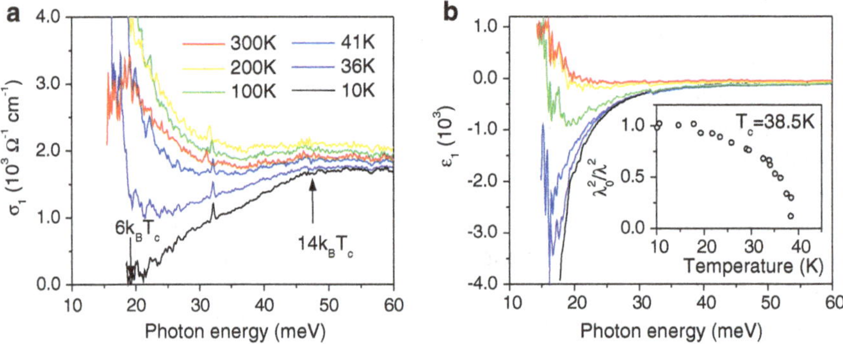

Fig. 4.6 Real part of the optical conductivity (**a**) and dielectric function (**b**) in the far-infrared spectral region. Two characteristic superconducting energy scales are present: $6k_BT_c$ and $14\,k_BT_c$. (Inset) Temperature dependence of the normalized inverse penetration depth obtained via a KK consistency check. Reprinted figure with permission from Ref. [39]. Copyright (2011) by the American Physical Society

with other measurements [18]. Application of a KK consistency check (as described in Sect. 3.5) to quantify the spectral weight contained in the extrapolation region below 10 meV allows one to rather accurately determine the temperature dependence of the London penetration depth. The resulting profile is shown in the inset of Fig. 4.6b for $1/\lambda_L(T)^2$ normalized to its value at 10 K.

At energies close to the optical superconducting gap $2\Delta_A \approx 20$ meV one of the directly measured ellipsometric angles $\Psi(\omega)$ approaches its critical value of 45° at the superconducting transition, which implies that the reflectivity of the sample approaches unity and its optical conductivity $\sigma_1(\omega)$ is close to zero. Remarkably, Fig. 4.6a shows a quasilinear dependence of $\sigma_1(\omega)$ in the superconducting state from $2\Delta_A$ to as high as $14k_BT_c$, in a stark contrast to the electron-doped 122 compounds [50–52]. In the latter the optical conductivity at $2\Delta_A$ decreases abruptly upon cooling below T_c, but only a weak superconductivity-induced modification is observed at higher energies. The quasilinear behavior in BKFA cannot be reconciled with the widely used for the pnictides Mattis-Bardeen theory [53], a weak-coupling extension of the BCS theory to finite impurity scattering. As all optimally-doped 122 pnictide superconductors appear to be in the strong-coupling regime, the Eliashberg theory [54] has to be used in order to obtain an adequate description of the optical properties.

4.2.4 Interband Optical Transitions

In order to analyze the free-charge-carrier response, the contribution of all interband transitions needs to be eliminated from the experimentally obtained optical conductivity. The total spectral weight of these transitions gives rise to a non-vanishing contribution $\Delta\varepsilon_{tot}$ to the real part of the far-infrared dielectric function $\varepsilon_1(\omega)$. In the iron pnictides $\Delta\varepsilon_{tot}$ has a particularly large value due to the high polarizability of the Fe–As bonds [1].

The full complex dielectric function $\varepsilon(\omega) = \varepsilon_1(\omega)+i\varepsilon_2(\omega)$ of $Ba_{0.68}K_{0.32}Fe_2As_2$, obtained experimentally in the range from 12 meV to 6.7 eV, was analyzed using

$$\varepsilon^{\text{inter}}(\omega) = \varepsilon_\infty + \sum_{j=1}^{n} \frac{\Delta\varepsilon_j\omega_{0j}^2}{(\omega_{0j}^2 - \omega^2) - i\Gamma_j\omega}, \tag{4.7}$$

where ε_∞ is the core contribution to the dielectric function and $\Delta\varepsilon_j$, ω_{0j}, Γ_j are the static permittivity contribution, center frequency and the width of the Lorentzian oscillators used to model the interband transitions, respectively. These parameters were obtained by a simultaneous fit of both the real and imaginary part of the dielectric function and are listed in Table 4.1. The remaining free-charge-carrier response will be analyzed in the following sections.

The results of such analysis are presented in Fig. 4.7 for $Ba_{0.68}K_{0.32}Fe_2As_2$ at 10 K (blue line—experimental data, black lines—separate Lorentz contributions).

Table 4.1 Parameters of the Lorentzian terms in Eq. (4.7) obtained from the analysis of the interband transitions in the optimally-doped BKFA

j	$\Delta\varepsilon_j$	ω_{0j}, cm^{-1}	Γ_j, cm^{-1}
1	44.37	4784.4	10243.0
2	5.67	6703.3	8557.4
3	1.10	13977.1	8914.0
4	7.31	20848.9	30158.4
5	0.473	23182.8	10268.9
6	2.54	58812.7	38689.8

Separate terms are shown in Fig. 4.7 as solid black lines

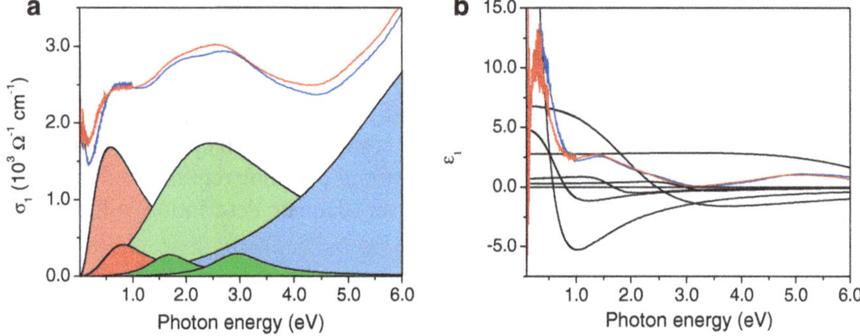

Fig. 4.7 Real part of the optical conductivity (**a**) and dielectric function (**b**) of $Ba_{0.68}K_{0.32}Fe_2As_2$ at 300 K (*red line*) and 10 K (*blue line*). Interband transitions inferred from the dispersion analysis (*solid black lines*). Reprinted figure with permission from Ref. [39]. Copyright (2011) by the American Physical Society

To display the scale of the temperature-induced variation of the interband transitions the experimental spectrum at 300 K is also shown (red line). The lowest interband transition in this material lies around 0.5 eV and significantly contributes to the optical polarizability of the system, as will be discussed in the next section. An unscreened bare plasma frequency of 1.6 eV at 41 K was consistently obtained from the spectral weight of the itinerant response $\mathrm{SW}_{\mathrm{it}} = \int_0^\infty \sigma_1^{\mathrm{it}}(\omega)d\omega$ as $\omega_{\mathrm{pl}} = \sqrt{8\mathrm{SW}_{\mathrm{it}}}$ and a simultaneous fit of the real and imaginary parts of the dielectric function at high energies.

4.2.5 Extended Drude Model

Signatures of a boson pairing mediator of the Eliashberg theory come from a qualitative analysis of the optical conductivity within the extended Drude model. It implies that the optical scattering rate is related to the far-infrared optical response as $\gamma(\omega) = \mathrm{Re}[\omega_{\mathrm{pl}}^2/4\pi\sigma^{\mathrm{it}}(\omega)]$, where ω_{pl} is the plasma frequency of free charge

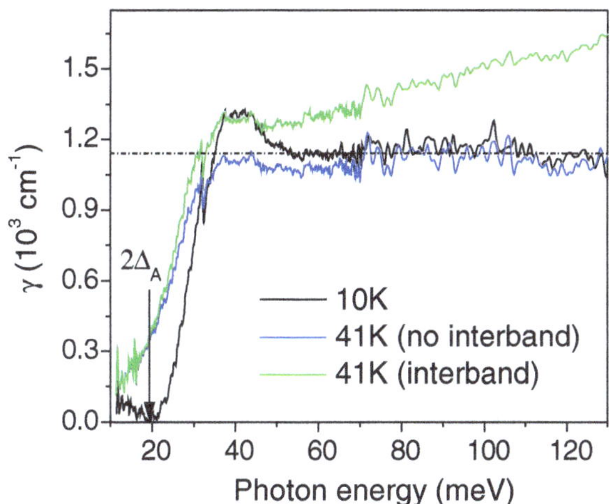

Fig. 4.8 Optical scattering rate obtained from the experimental data at 41 and 10 K within the extended Drude model, with the contribution of the interband transitions subtracted (*blue* and *black lines*, respectively) and at 41 K without subtraction (*green line*). *Dash-dotted line* indicates the saturation level of the high-energy optical scattering rate. Reprinted figure with permission from Ref. [39]. Copyright (2011) by the American Physical Society

carriers. The superscript 'it' stands for 'itinerant' and implies that the contribution of all interband transitions has been subtracted from the experimentally obtained optical conductivity as described in the previous section. The optical scattering rate shows clear evidence for an intermediate boson irrespective of complications due to the multiband character of the compound. Figure 4.8 plots $\gamma(\omega)$ of BKFA at 41 K (blue line) and 10 K (black line) for $\omega_{pl} = 1.6\,\text{eV}$ with all interband transitions subtracted in both cases (the itinerant spectral weight SW_{it} corresponds to the blue shaded area in Fig. 4.9a). In the superconducting state, no scattering is expected up to photon energies exceeding the binding energy of the Cooper pairs. Thus the onset of the optical scattering rate in Fig. 4.8 marks the optical energy gap $2\Delta_A = 20\,\text{meV}$. Saturation of $\gamma(\omega > 50\,\text{meV})$ at $1100\,\text{cm}^{-1}$ indicates that the boson spectral function is contained well below 50 meV [55]. It is important to emphasize that due to the multiband character of the iron pnictides the analysis of the optical scattering rate in the framework of a single-band Eliashberg theory is potentially misleading. Moreover, also shown in Fig. 4.8 is a spectrum that directly results from the experimental data, without accounting for the interband transitions. It becomes clear that an increase in the scattering rate at higher energies that might be ascribed to strong electron correlations can result from an unsubtracted contribution of the interband transitions to the complex optical conductivity. This is especially important in the iron pnictides since the lowest lying interband transition at about 0.5 eV contributes to the anomalously large value of the low-energy dielectric permittivity $\Delta\varepsilon_{tot} = \sum_{j=1}^{n} \Delta\varepsilon_j \approx 60$ (see Table 4.1). This value is clearly seen in Fig. 4.9b as the zero-frequency limit of $\varepsilon_1^{inter}(\omega)$ determined by means of a dispersion analysis. It is also fully consistent with the bare plasma frequency of 1.6 eV and the zero-crossing in $\varepsilon_1(\omega)$ at 0.2 eV (blue line in Fig. 4.9b). Such $\Delta\varepsilon_{tot}$ is an order of magnitude larger than in any other high-temperature superconductor (e. g. ≈5 in the cuprates [10]). Recently, a similarly high

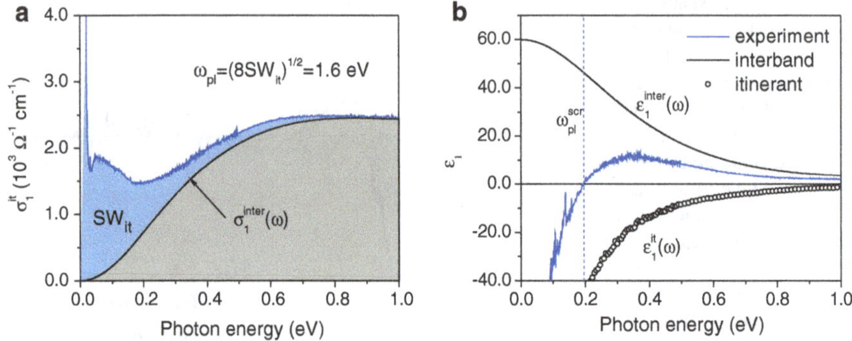

Fig. 4.9 **a** Real part of the optical conductivity at 41 K (*blue line*). The contribution of itinerant charge carriers (*blue area*) is obtained by subtracting all interband transitions $\sigma_1^{\text{inter}}(\omega)$ (*gray area*) from the optical response. **b** Real part of the dielectric function at 41 K (*blue line*). The free-charge-carrier response $\varepsilon_1^{\text{it}}(\omega)$ (*open circles*) is obtained by eliminating all interband transitions $\varepsilon_1^{\text{inter}}(\omega)$ (*black solid line*). The *blue dashed line* indicates the screened plasma frequency at 41 K. Reprinted figure with permission from Ref. [39]. Copyright (2011) by the American Physical Society

value in a conventional superconductor was inferred from reflectivity measurements on elemental bismuth [56].

4.2.6 *Effective Two-Band Model for Iron Pnictides*

Four-Band Eliashberg Theory

To determine the microscopic origin of the high-energy anomaly $2\Delta_\text{A} < \hbar\omega < 14k_\text{B}T_\text{c}$ in the real part of the optical conductivity in Fig. 4.6a we use the four-band Eliashberg theory [54, 57, 58] that proved successful in explaining thermodynamical data obtained on the same compound [17].

The main ingredient of the theory is the total spectral function of the electron-boson interaction $B(\omega)$ (Eliashberg function; analogous to that of the electron-phonon interaction $\alpha^2F(\omega)$). In a four band system it can be decomposed into 16 functions $B(\omega)_{ij}$, where i and j label the four Fermi surface sheets ($i, j = \overline{1,4}$). The *standard* Eliashberg functions determine superconducting and thermodynamical properties such as the superconducting transition temperature and gaps, electronic specific heat, de Haas-van Alphen mass renormalizations etc., and are defined as

$$B(\omega)_{ij} = \frac{1}{N_i} \sum_{\mathbf{k},\mathbf{k}',\nu} \left| g_{\mathbf{k},\mathbf{k}'}^{ij,\nu} \right|^2 \delta(\varepsilon_\mathbf{k}^i)\delta(\varepsilon_{\mathbf{k}'}^j)\delta(\omega - \omega_{\mathbf{k}-\mathbf{k}'}^\nu),$$

where N_i is the partial DOS per spin on the ith sheet of the Fermi surface, and $g_{\mathbf{k},\mathbf{k}'}^{ij,\nu}$ is the matrix element of the electron-boson interaction.

Transport and electrodynamical properties are defined by 16 *transport* Eliashberg functions (which enter the Boltzmann kinetic equation)

$$B(\omega)_{\mathrm{tr},ij}^{\alpha\beta} = \frac{1}{2N_i \left\langle v_{\mathrm{F},i}^{\alpha\,2} \right\rangle} \sum_{\mathbf{k},\mathbf{k}',\nu} \left| g_{\mathbf{k}\mathbf{k}'}^{ij\nu} \right|^2$$
$$\times \left(v_{\mathrm{F},i}^{\alpha}(\mathbf{k}) - v_{\mathrm{F},j}^{\beta}(\mathbf{k}')\right)^2 \delta(\varepsilon_{\mathbf{k}}^{i})\delta(\varepsilon_{\mathbf{k}'}^{j})\delta(\omega - \omega_{\mathbf{k}-\mathbf{k}'}^{\nu}),$$

where $v_{\mathrm{F},i}^{\alpha}$ is the α-th Cartesian component of the Fermi velocity on Fermi surface i. The average Fermi velocity is related to the plasma frequency by the standard expression $\omega_{\mathrm{pl},i}^2 = 8\pi e^2 N_i \left\langle v_{\mathrm{F},i}^2 \right\rangle = 8\pi e^2 \sum_{\mathbf{k}} v_{\mathrm{F},i}^2(\mathbf{k})\delta(\varepsilon_{\mathbf{k}}^i)$. All Eliashberg functions satisfy the symmetry relations $M_i B_{ij} = M_j B_{ji}$, where $M_i = N_i$ and $M_i = \omega_{\mathrm{pl},i}^2$ for the standard and transport Eliashberg functions, respectively.

As a starting point we consider the 4-band model based on the band-structure calculations with two hole bands and two electron bands crossing the Fermi level that has proven successful in accounting for the thermodynamical properties of BKFA [17]. We use the same input parameters, namely, the densities of states $N_1 = 22$, $N_2 = 25$, and $N_3 = N_4 = 7$ states per Rydberg and per unit cell, the first two having a hole while the other two an electron character. The main input, the spectral function of the intermediate boson, was taken the same as in Fig. S4 of Ref. [17] in the form of a spin-fluctuation spectrum $\tilde{B}_{ij}(\Omega) = \lambda_{ij} f(\Omega/\Omega_{sf})$ with a linear ω dependence at low frequencies. Here λ_{ij} is the coupling constant pairing band i with band j and Ω_{SF} is a characteristic spin-fluctuation frequency, the values of which correspond to those in Table SII of Ref. [17]: $\Omega_{sf} = 13\,\mathrm{meV}$ and

$$\lambda_{ij} = \begin{pmatrix} 0.2 & 0 & -1.7 & -1.7 \\ 0 & 0.2 & -0.25 & -0.25 \\ -5.34 & -0.89 & 0.2 & 0 \\ -5.34 & -0.89 & 0 & 0.2 \end{pmatrix}. \tag{4.8}$$

The negative elements correspond to *interband* hole-electron repulsion, while the positive—to *intraband* attraction.

The 4×4 matrix of the coupling constants and 4 densities of states characterizing this model are highly constrained by thermodynamic, transport, and photoemission data [17, 36, 43, 44, 46, 47], and the same set of parameters is used here. In principle, an additional set of 4 plasma frequencies and a 4×4 matrix of the intraband/interband impurity scattering rates has to be taken into account to describe the optical response. However, this parameter set can be strongly reduced based on the following considerations.

Role of Impurities and Defects

Both normal and superconducting properties of a multiband superconductor significantly depend on impurity scattering. Unlike in conventional superconductivity, one has to distinguish between intraband impurity scattering, which does not add any new physics (in the Born approximation) compared with single-band superconductivity, and *interband* scattering, which in many cases has an effect comparable to the pair-breaking effect of magnetic impurities (or of nonmagnetic impurities in superconductors with p- or d-wave pairing) [59]. In this regard, the fact that no strong correlation has been observed between the residual resistivity (which indirectly characterizes impurity scattering) and the critical temperature T_c of the (nearly) optimally electron-doped BKFA (see Table 4.2) indicates that the concentration of *interband* impurities in the Born limit is very small. Thus one only needs to estimate the intraband scattering rates.

Reduction to a Two-Band Model

A substantial simplification can further be made by projecting the four-band model onto an effective two band model, motivated by the observation of two distinct groups of superconducting energy gaps in a variety of experiments [17, 36, 43, 44, 46, 47]. These gaps can be identified as a single gap Δ_B on the outer holelike Fermi surface and a group of three gaps of magnitude $\sim\Delta_A$ on the inner holelike and the two electronlike Fermi surfaces.

In general, the superconducting order parameters are a solution of a linear system of equations

$$e_i = \sum_{j=1}^{4} B_{ij}(\omega)e_j. \tag{4.9}$$

The Eliashberg functions $B_{ij}(\omega)$ satisfy the symmetry relations

$$N_i B_{ij}(\omega) = N_j B_{ji}(\omega) \tag{4.10}$$

and, therefore, can be represented in the form $B_{ij}(\omega) = U_{ij}(\omega)N_j$, where U_{ij} is a symmetrical matrix. Further, we can construct a functional

Table 4.2 Superconducting transition temperature and residual resistivity ρ_{40K} of (nearly) optimally hole-doped BKFA

T_c, K	Residual resistivity, mΩ cm	Reference
38.5	0.04	[17]
38	0.075	[60]
38	0.1	[61]
36.5	0.055	[62]

$$F\{e_i\} = \sum_{j=1}^{4} N_j e_j^2 - \sum_{i,j=1}^{4} N_i e_i U_{ij} N_j e_j, \tag{4.11}$$

the minimization of which with respect to e_i results in Eq. (4.9). As mentioned above, BKFA has three gaps with very close absolute values (the hole gap has the opposite sign with respect to the other two). Minimizing the functional in Eq. (4.11) subject to the additional constraints $e_3 = e_4 = -e_1 = \Delta_A$ and $e_2 = -\Delta_B$ one finds

$$\begin{pmatrix} \Delta_A \\ \Delta_B \end{pmatrix} = \begin{pmatrix} \lambda_{AA} & \lambda_{AB} \\ \lambda_{BA} & \lambda_{BB} \end{pmatrix} \begin{pmatrix} \Delta_A \\ \Delta_B \end{pmatrix}, \tag{4.12}$$

where the matrix elements satisfy

$$\begin{aligned} \lambda_{AA} &= \frac{N_1 (\lambda_{11} - 2\lambda_{13} - 2\lambda_{14}) + N_3\lambda_{33} + N_4\lambda_44}{N_1 + N_3 + N_4}, \\ \lambda_{AB} &= \frac{N_2 (\lambda_{23} + \lambda_{24})}{N_1 + N_3 + N_4}, \\ \lambda_{BA} &= \lambda_{23} + \lambda_{24}, \\ \lambda_{BB} &= \lambda_{22}. \end{aligned}$$

Given a boson spectrum centered at 13 meV (for details see supplementary online material in Ref. [17]), consistent with the energy of the spin resonance excitation in this compound[3] [63] and assuming the matrix elements in Eq. (4.8), one obtains the following two-band model coupling matrix:

$$\lambda_{IJ} = \begin{pmatrix} 4.36 & -0.35 \\ -0.5 & 0.2 \end{pmatrix}, \quad I, J = \{A, B\}. \tag{4.13}$$

The partial DOSes of effective band A and band B on the Fermi level are

$$\begin{aligned} N_A &= N_1 + N_3 + N_4 = 36 \text{ states/(Rydberg u.c.)}, \\ N_B &= N_2 = 25 \text{ states/(Rydberg u.c.)}, \end{aligned} \tag{4.14}$$

where ‘u.c.’ stands for ‘unit cell’. The first effective intraband coupling constant is an order of magnitude larger than predicted for the intraband electron-phonon coupling [15]. It does not, however, have a direct physical meaning by itself but rather incorporates the contributions of *three* different bands. We reiterate that the coupling matrix has been inferred from prior measurements. This way only two intraband impurity scattering rates enter as free parameters of the theory in addition to the plasma frequencies of the bands.

[3] The feedback of superconductivity on the spin fluctuation spectrum observed by neutron scattering [63, 64] was not considered in our model calculation.

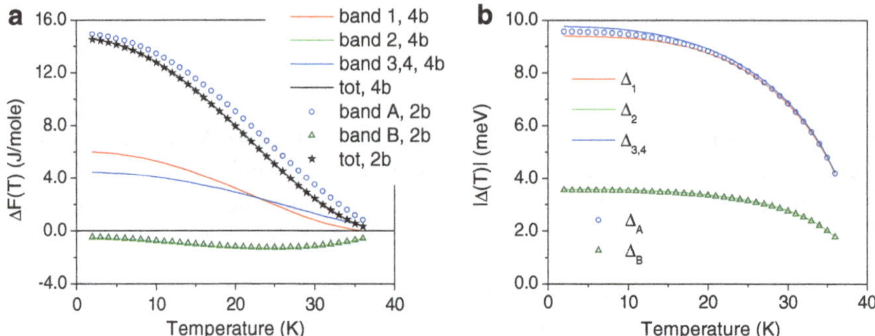

Fig. 4.10 Temperature dependence of the free energy (**a**) and superconducting gaps (**b**) in the four-band (*lines*) and reduced two-band (*symbol*) models, with coupling matrices Eqs. (4.8, 4.13), respectively. Reprinted figure with permission from Ref. [39]. Copyright (2011) by the American Physical Society

The effective two-band model closely reproduces all the predictions of the four-band model such as the superconducting transition temperature $T_c = 38.4$ K, superconducting gaps $\Delta_A = 9.7$ meV and $\Delta_B = 3.7$ meV, free energy and the dependence of the superconducting gaps on temperature, as shown in Fig. 4.10a, b, respectively. The calculated DOSes N_A and N_B are very similar, in accordance with the partial Sommerfeld constants obtained in the treatment of the specific heat data in a phenomenological two-band α-model [17]: $\gamma_A \simeq \gamma_B$, where the Sommerfeld constant γ is related to the DOS per spin at the Fermi energy $N(0)$ via $\gamma = (1/3)\pi^2 k_B^2 N(0)$.

4.2.7 Comparison with Experiment

In our calculation we consider two clean bands with $\gamma_A = \gamma_B = 1\,\text{cm}^{-1}$. As the bare plasma frequencies of all bands are similar it follows that the spectral weight of band A has to be much larger than that of band B. Assigning 80% of the spectral weight to the effective band we obtain the results presented as solid lines in Fig. 4.11. The high-energy anomaly at $14k_B T_c$ is naturally captured by the model without resorting to additional gaps. By varying the effective coupling strength in our calculations we found that this energy could be assigned to $4\Delta_A(0) + \Omega$, where Ω is the peak frequency of the boson spectrum, as shown in Fig. 4.12. This calculation also accounts for the fact that only the biggest superconducting gap is visible in the optical response of BKFA due to the small contribution of band B (20 % of the spectral weight) to the total optical conductivity. This leads to two possible levels of the impurity scattering rate of band B, which has to be either very small $\gamma_B \approx 1\,\text{cm}^{-1}$ or very large at about $1000\,\text{cm}^{-1}$. The latter value provides a better description of the optical scattering rate (see interactive simulation in Ref. [66]). However, such a large disparity between the charge carriers is hard to reconcile with Hall and de

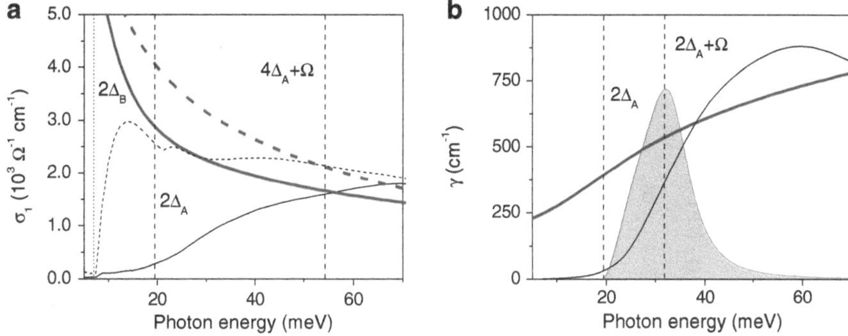

Fig. 4.11 **a** Real part of the far-infrared conductivity obtained within the two-band Eliashberg theory (see text) at 40 K (*blue thick lines*) and 10 K (*black thin lines*) in the clean limit $\gamma_A = \gamma_B = 1\,\mathrm{cm}^{-1}$ (*solid lines*) and dirty limit $\gamma_A = \gamma_B = 200\,\mathrm{cm}^{-1}$ (*dashed lines*). **b** Optical scattering rate in the clean limit from the same model. The *gray area* shows the normalized boson spectral function $B(\omega)$ used in the calculation, displaced from zero by $2\Delta_A$ to assist interpretation in the superconducting state. Reprinted figure with permission from Ref. [39]. Copyright (2011) by the American Physical Society

Haas–van Alphen experiments, which imply that the impurity scattering rate of holes is no more than one order of magnitude higher than that of electrons [67, 68]. This residual uncertainty notwithstanding, our results show that the impurity scattering rate of band A must be very small, because the energy $4\Delta_A(0) + \Omega$ is no longer discernible in the simulated spectra when γ_A increases (interactive simulation in Ref. [66]). Thus the region of the linear increase of $\sigma_1(\omega)$ can only be observed in a very clean material.

The same reduced two-band model can be applied to the case of $BaFe_{1.85}Co_{0.15}As_2$ (BFCA). In this compound the spin-resonance excitation occurs at a very similar energy of 10 meV [64]. A boson spectrum centered at this energy is also consistent with Andreev-reflection measurements [69]. Recently, a comprehensive specific-heat study of this compound at different Co-doping levels has been carried out [70]. The analysis of the experimental data in the framework of a two-band α-model indicates that the largest gap develops in the band with the largest electronic DOS, providing further evidence that several bands contribute to the strongly-coupled band in the reduced two-band model. Figure 4.11a (dashed lines) shows that a calculation within the same reduced two-band model qualitatively reproduces the far-infrared optical conductivity of BFCA [50–52] when both bands are assumed to be dirty with $\gamma_A = \gamma_B = 200\,\mathrm{cm}^{-1}$ and a redistribution of the spectral weight between the bands is taken into account as $\omega^2_{\mathrm{pl,A}} \approx \omega^2_{\mathrm{pl,B}}$. The model captures two prominent superconductivity-induced anomalies clearly observed in experiments: the steep onset of absorption at the value of the small gap $2\Delta_B$ and the weaker superconductivity-induced changes of the optical conductivity extending up to $18k_BT_c$. The redistribution of the spectral weight between the bands in BFCA compared to BKFA implied by our analysis is justified by doping with different

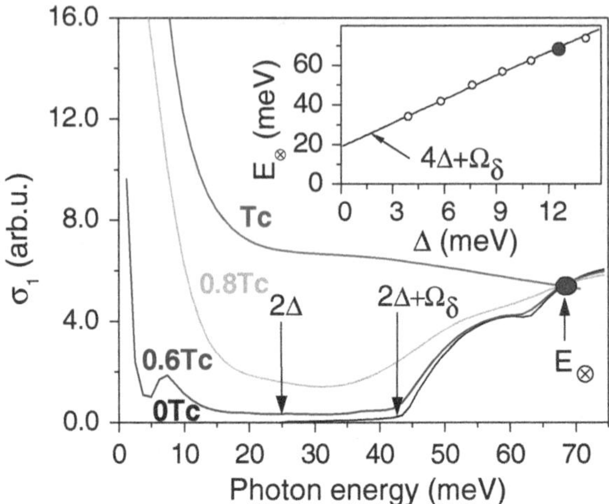

Fig. 4.12 Real part of the optical conductivity at several temperatures obtained within the single-band Eliashberg theory, with the spectral shape of the Eliashberg function taken as a δ-function and the coupling constant varied to change the value of the superconducting gap. Due to the singular shape of the Eliashberg function all characteristic energy features in the optical conductivity are clearly visible, such as the optical excitation gap 2Δ and the Holstein excitation threshold $2\Delta + \Omega$. Absorption below 2Δ is energetically forbidden because there is not enough photon energy to break up a Cooper pair, whereas absorption above it is kinematically forbidden in the clean case (with no impurity scattering) due to the conservation of momentum [65]. In our case small impurity scattering gives rise to very weak absorption above 2Δ. Excitation of quasiparticles becomes fully allowed above $2\Delta + \Omega$ where the photon energy is sufficient to break up a Cooper pair and the excess momentum is carried away by a quantum of the mediating boson with energy Ω. (Inset) The dependence of the intersection energy $E_{\otimes}$ in the real part of the optical conductivity on the coupling constant via the superconducting gap plotted versus the latter. The *filled blue circle* represents the case displayed in the main panel

carriers in the two compounds, while the large difference in their impurity scattering rates is a natural consequence of the difference in the doping mechanism by chemical substitution, which directly affects the FeAs layers in BFCA, but not in BKFA.

4.2.8 Conclusions

In summary, a qualitative description of superconductivity-induced optical anomalies in the far-infrared optical conductivity of $Ba_{0.68}K_{0.32}Fe_2As_2$ is obtained in the framework of an effective two-band Eliashberg theory with a strong coupling to spin fluctuations reduced from its four-band counterpart. The linear increase of absorption above the larger superconducting gap can only be observed when the effective band is extremely clean. The same model in the dirty limit gives a good qualitative explanation of the optical conductivity of the optimally electron-doped BFCA consistently in the strong-coupling regime. Our approach provides a universal description of the far-infrared conductivity of the 122 iron pnictides and clearly outlines the limitations of the widely-used for analysis of essentially all iron-based superconductors Mattis-Bardeen theory. It is indispensable for the correct determination of the number and magnitudes of the superconducting gaps in clean materials.

4.3 Optical Conductivity of Superconducting $Rb_2Fe_4Se_5$

4.3.1 Introduction

In the family of the iron-pnictide/chalcogenide superconductors most research effort has so far been applied to the so-called 122 compounds with Fe–As conducting planes due to the high quality and large size of the single crystals available.[4] They bear all hallmarks of this new class of superconductors such as the itinerant antiferromagnetic ground state of the parent compounds, multiple bands crossing the Fermi level, superconducting transition temperatures up to 40 K, and a resonance peak in the INS signal at the (0, 0.5) or (0.5, 0) **Q**-vector in the superconducting state [42], suggesting novel superconductivity with the s-wave symmetry and a sign change of the order parameter between the hole and electron Fermi pockets [14]. Throughout the phase diagram these compounds are metals with a plasma frequency $\omega_{pl} \approx 1.6\,eV$ close to the optimal doping level [1, 28, 72]. In the superconducting state the corresponding optical conductivity is fully suppressed below 2Δ due to the formation of a superconducting condensate with a London penetration depth of $\lambda_L \approx 220\,nm$ [18, 39, 51].

Recently, iron selenide compounds have been synthesized in this class of superconductors [73–76]. They were first believed to crystallize in the same I4/mmm symmetry of $ThCr_2Si_2$ type but soon it became clear that there is an inherent iron-vacancy order present in these materials with a chiral $\sqrt{5} \times \sqrt{5} \times 1$ superstructure, which reduces the symmetry to I4/m and makes it more appropriate to classify these materials into the 245 stoichiometry [77]. The Fe-vacancy and antiferromagnetic orders occur at rather high transition temperatures of 400–550 K. Neutron-scattering studies showed that these compounds possess a magnetic moment on the iron atoms of about $3.3\,\mu_B$ [78], which is unusually large for the iron pnictides. At the same time a resonance peak has been observed by INS below $T_c \approx 32$ K at an energy of $\hbar\omega_{res} = 14\,meV$ and the **Q**-vector (0.5, 0.25, 0.5) in the unfolded Fe-sublattice notation [79], which is also unprecedented for the iron pnictides. It is still under debate how superconductivity with such a high transition temperature can exist on such a strong magnetic background although there are some indications of inherent phase separation in the iron chalcogenides [80–83]. A further complication arises from the fact that, unlike their 122 counterparts, the 245 iron selenides show a semiconducting optical response and no free charge carrier conductivity has been reported so far [84].

[4] The discussion in this section is largely based on Ref. [71].

In this work we provide first direct evidence for a free-charge-carrier contribution to the optical response in RFS. We show that the charge-carrier density is small, with $\omega_{\mathrm{pl}} \approx 100\,\mathrm{meV}$. This metallic response experiences a weak modification upon cooling into the superconducting state. This evidence together with the results of resistivity, magnetization, specific-heat measurements [85], and Mössbauer spectroscopy [86] indicates that in this compound superconductivity and magnetism coexist as two separate phases.

4.3.2 Experimental Details

Optimally-doped superconducting RFS single crystals were grown by the Bridgman method (batch BR16 in Ref. [85]). From DC resistivity, magnetization and specific-heat measurements we obtained $T_c \approx 32\,\mathrm{K}$. Sample cleaving and handling was carried out in argon atmosphere at all times prior to every optical measurement. The complex dielectric function $\varepsilon(\omega) = \varepsilon_1(\omega) + i\varepsilon_2(\omega) = 1 + 4\pi i \sigma(\omega)/\omega$, where $\sigma(\omega)$ is the complex optical conductivity, was obtained in the 0.01–6.5 eV range using broadband ellipsometry, as described in Ref. [49]. Time-domain THz transmission measurements were carried out in the 1–10 meV spectral range using a TPS spectra 3000 spectrometer with f/2 focusing optics (TeraView Ltd.). The samples were mounted on a square copper sample holder of 17 mm edge length and an aperture of 1.5 mm in diameter. The sample's edges were covered with silver paint to avoid radiation leakage. The transmitted intensity and phase, obtained via the Fourier transform of the time-domain spectra, were used to directly calculate the complex dielectric function. The far-infrared optical response was measured at the infrared beamline of the ANKA synchrotron light source at Karlsruhe Institute of Technology, Germany.

4.3.3 Broadband Spectroscopic Ellipsometry

The imaginary and real part of the complex dielectric function in the $0.01-6.5\,\mathrm{eV}$ spectral range are shown in Fig. 4.13a, b, respectively. Down to 10 meV the sample does not reveal any metallic behavior as is evident from ε_1, which remains positive at all temperatures. It also displays several infrared-active optical phonons, similar to those previously observed in the far-infrared optical response of semiconducting $K_2Fe_4Se_5$ (KFS) [84]. More infrared-active phonons observed than allowed by the tetragonal symmetry of the 122 unit cell supports the reduction of the Brillouin zone due to the ordering of iron vacancies. Throughout the whole far-infrared $10-100\,\mathrm{meV}$ spectral range we find a rapid increase of the electronic background in $\sigma_1(\omega)$ between 200 and 100 K with a concomitant decrease in ε_1. At higher frequencies the optical response features the onset of interband transitions around 0.25 eV. It is followed by an absorption edge at $\approx$0.45 eV formed by direct interband transitions, as shown in the inset of Fig. 4.13a. Unlike in the 122 compounds [28], the

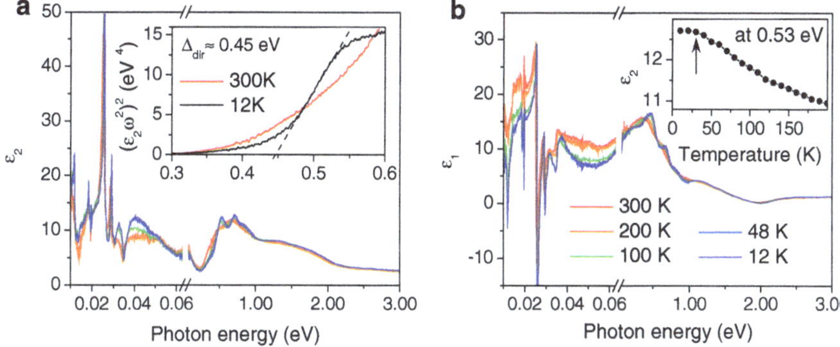

Fig. 4.13 Imaginary (**a**) and real (**b**) part of the dielectric function of RFS in the 0.01–6.5 eV spectral range at different temperatures. (Inset in **a**) Plot of $(\varepsilon_2(\omega)\omega^2)^2$ near the absorption edge. The intersection of the *dashed line* with the energy axis defines the direct energy gap $\Delta_{dir} = 0.45$ eV at 12 K. (Inset in **b**) Temperature dependence of ϵ_2 at 0.53 eV. Reprinted figure with permission from Ref. [71]. Copyright (2012) by the American Physical Society

lowest-lying absorption band peaked at about 0.6 eV reveals three separate contributions at low temperatures similar to the two contributions reported for the semiconducting KFS [84]. The inset in Fig. 4.13b shows the strong temperature dependence of one of these absorption bands in the magnetic state, which is linear in a broad temperature range. This behavior is fully consistent with the temperature dependence of the magnetic Bragg peak intensity, including the saturation at 30–40 K [87], which suggests a spin-controlled character of these interband transitions.

4.3.4 *Terahertz Time-Domain Transmission*

As the compound remains transparent down to 10 meV, a pathway opens to apply time-domain transmission spectroscopy to obtain the complex dielectric function of RFS also at THz frequencies, should sufficiently thin samples be obtained. It was indeed possible to achieve sizable transmission in cleaved flakes of RFS about 25-μm thick, with terraces less than 100- nm high over $50 \times 50\,\mu m^2$ areas. Figures 4.14a, b show ε_2 and ε_1, respectively, obtained in the 1–10 meV spectral range using this technique for temperatures $4 \leq T \leq 300$ K, along with the optical response of a 100-μm thick insulating sample BR17 grown by the same method [85]. In the time-domain signal of the insulating sample the first reflection is clearly visible about 3 ps after the main pulse (see the inset) and results in a superposed interference pattern on its complex dielectric constant. This sample shows a typical frequency-independent insulating response even at 12 K (magenta lines in Fig. 4.14). Typical electric-field transients obtained on these samples are shown in the inset. The superconducting sample shows much stronger absorption (solid red line) than the insulating one (solid blue line) due

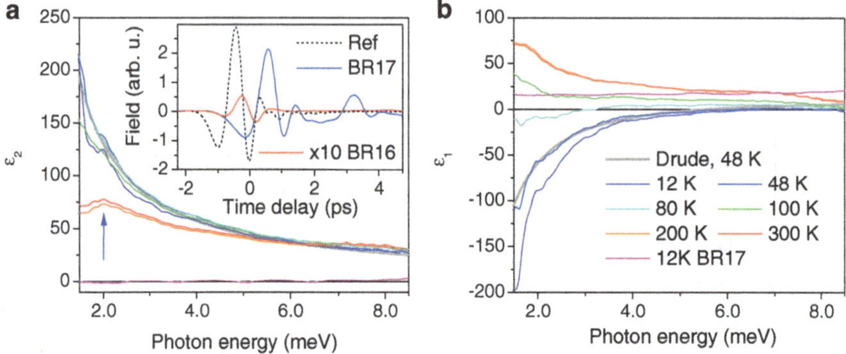

Fig. 4.14 Imaginary (**a**) and real (**b**) part of the dielectric function of RFS in THz spectral region obtained via the time-domain transmission spectroscopy. *Blue arrow marks* a low-energy electronic excitation. *Thick gray lines* show a Drude fit to $\varepsilon_1(\omega)$ and $\varepsilon_2(\omega)$ as described in the text. (Inset) Electric field transients transmitted through a 25-μm thick superconducting sample (multiplied by 10, *red line*) and 100-μm thick transparent insulating sample (*blue line*). The *dashed line* plots the reference signal. Reprinted figure with permission from Ref. [71]. Copyright (2012) by the American Physical Society

to a high level of the electronic background in $\varepsilon_2(\omega)$. At room temperature and down to 100 K the former remains semiconducting, with $\varepsilon_1(\omega)$ positive in the whole spectral range. However, unlike the insulating sample (thick magenta line in Fig. 4.14b), it exhibits an upturn at lowest energies, which indicates a low-energy electronic mode peaked at 2 meV (blue arrow in Fig. 4.14a). Below 100 K a clear metallic response with negative $\varepsilon_1(\omega)$ rapidly develops. Already at 80 K the zero-crossing in $\varepsilon_1(\omega)$ corresponds to a screened plasma frequency of 3 meV, which reaches 6.5 meV as the temperature is lowered further. The metallic response can be fitted by two Drude terms with $\omega_{pl,1} \approx 20$ meV, $\gamma_1 \approx 1$ meV and $\omega_{pl,2} \approx 95$ meV, $\gamma_2 \approx 40$ meV for the plasma frequencies and scattering rates of the narrow and broad components at 48 K, respectively (thick gray lines in Fig. 4.14). The total plasma frequency of the free-charge-carrier response is given by $\omega_{pl} = \sqrt{\omega_{pl,1}^2 + \omega_{pl,2}^2} \approx 100$ meV. The observed crossover from semiconducting to metallic behavior with decreasing temperature below 100 K is in full agreement with the temperature dependence of the dc resistivity [85]. The spectral weight of the low-energy electronic mode in the semiconducting state amounts to about 4 % of the total spectral weight of the free charge carriers and might originate from the narrow Drude component. This low-energy mode can represent a collective electronic excitation pinned by structural defects like iron vacancies.

The discovery of itinerant charge carriers in the optimally-doped RFS requires a more detailed study of its low-temperature optical response in the vicinity of the superconducting transition temperature. Difference spectra of the real part of the optical conductivity and the dielectric function are shown in Figs. 4.15a, b, respectively, for several temperatures between 12 and 60 K. The sample shows moderate

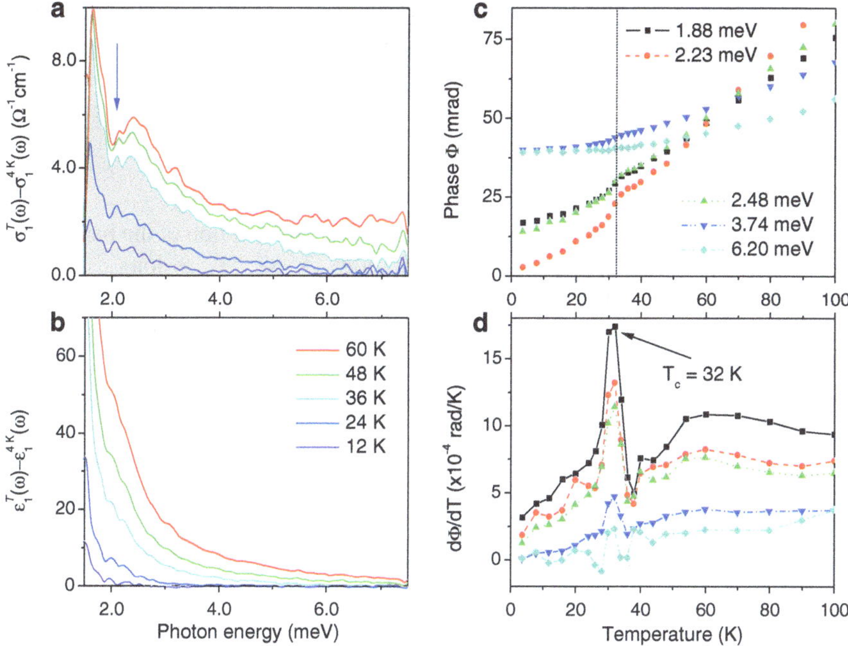

Fig. 4.15 Difference spectra $\sigma_1(T, \omega) - \sigma_1(4\ \mathrm{K}, \omega)$ (**a**) and $\varepsilon_1(T, \omega) - \varepsilon_1(4\ \mathrm{K}, \omega)$ (**b**) of a 25 μm superconducting sample. Blue arrow marks the same low-energy electronic excitation as in Fig. 4.14a. *Shaded area* indicates the spectral weight used for the estimation of the ω_{pl} of the superconducting condensate (see text). **c** Temperature dependence of the time-domain transmission phase and **d** its temperature derivative for several frequencies. Superconducting transition temperature is 32 K (*dashed vertical line*). Reprinted figure with permission from Ref. [71]. Copyright (2012) by the American Physical Society

changes in the normal state and a rapid decrease of the optical conductivity below T_c. The missing area between 36 and 4 K shown in grey in Fig. 4.15a gives rise to a characteristic $-1/\omega^2$ contribution to $\varepsilon_1(\omega)$. Using the KK consistency analysis of the relative changes in the complex dielectric function (see Sect. 3.5), we determine the superconducting plasma frequency $\omega_{\mathrm{pl}}^{\mathrm{SC}} \approx 10\,\mathrm{meV}$. Further evidence for the superconductivity-induced nature of these changes in the optical response comes from the temperature dependence of the transmission phase shown in Fig. 4.15c for several frequencies. At all frequencies up to 8 meV there is a kink at $T_c \approx 32\,\mathrm{K}$, which gets progressively smaller as the frequency increases. This effect is more obvious in the temperature derivative of the transmission phase shown for the same frequencies in Fig. 4.15d. In addition, we observe a double-peak structure around 2 meV (blue arrow in Fig. 4.15a; see also Fig. 4.14a), which is overwhelmed by the electronic background at 300 K but clearly stands out at lower temperatures due to reduced electron scattering and even persists in the superconducting state. This feature might have its origin in the electron and hole bound states induced by iron vacancies recently observed at similar energies in an STM/STS study on KFS in Ref. [82], which might serve as pinning centers for a collective electronic excitation.

4.3.5 LDA Calculations

It is known that LDA calculations provide an adequate description of the band structure and optical conductivity of the iron pnictides [1, 88, 89] as long as a moderate mass and bandwidth renormalization is taken into account. We compare the experimentally obtained $\sigma_1(\omega)$ with the theoretical prediction for the $Rb_2Fe_4Se_5$ compound. The calculation was performed for the experimental $\sqrt{5} \times \sqrt{5} \times 1$ superstructure [77] assuming the so-called block-checkerboard antiferromagnetic order of the Fe moments [90]. In Figs. 4.16a, b the experimentally obtained spectra of $\sigma_1(\omega)$ and $\varepsilon_1(\omega)$ for the 122 BKFA and 245 RFS systems are compared with the results

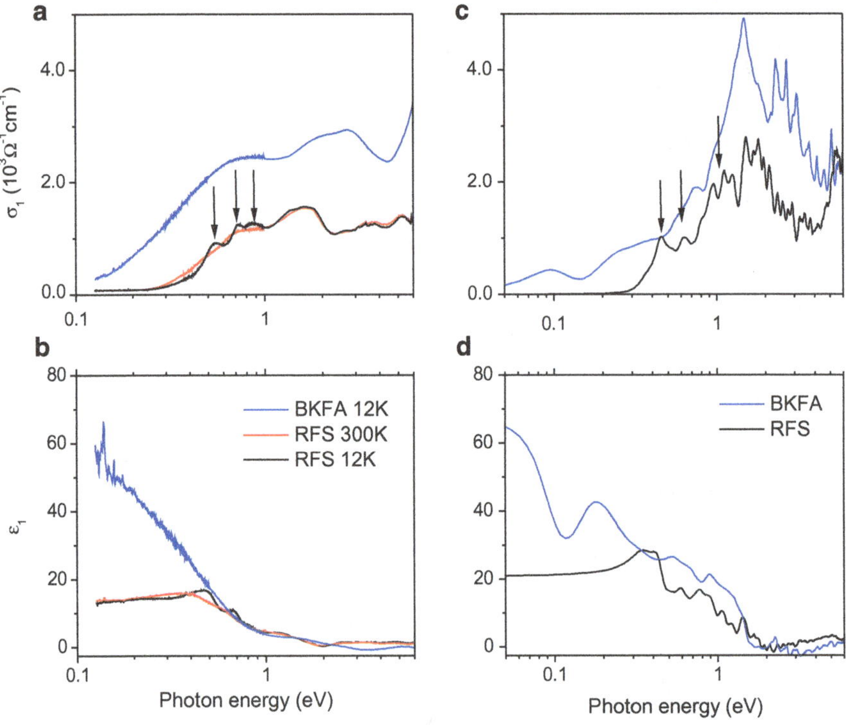

Fig. 4.16 **a, b** Comparison of the experimentally-obtained real part of the (**a**) optical conductivity and (**b**) dielectric function of optimally-doped BKFA at 12 K and RFS at 300 and 12 K. **c, d** Same as in *left panels* but as obtained in LDA calculations. The *arrows* in **c** and **d** indicate the narrow low-energy optical bands that can be resolved in RFS at low temperatures but not in BKFA. Reprinted figure with permission from Ref. [71]. Copyright (2012) by the American Physical Society

of LDA calculations (c) and (d), respectively. Already the direct comparison of the experimentally-obtained $\sigma_1(\omega)$ and $\varepsilon_1(\omega)$ of BKFA and RFS shows that the overall structure of the interband transitions in these two classes of superconductors is very similar apart from frequency shifts (blue and red lines in Fig. 4.16a, respectively; the overall shape of the interband optical conductivity of BKFA is virtually unchanged between 300 and 10 K, see Sect. 4.2). The most striking difference is the narrowing of the absorption bands in RFS around 0.6 eV at low temperatures uncovering three distinct components (black arrows in Fig. 4.16a). Similarly, the appearance of this fine structure can be observed in the LDA calculations, as shown in Fig. 4.16c for BKFA and RFS (blue and black lines, respectively, arrows indicate three possible contributions to the resolved fine structure of the 0.6 eV absorption band). It has already been shown in Sect. 4.2 by subtracting the itinerant contribution to the infrared optical response of BKFA that the low-frequency dielectric permittivity in this compound has an anomalously large value of about 60 most likely due to the high polarizability of the Fe–As bonds. In RFS, the very weak and narrow free charge carrier response allows for the determination of the low-frequency permittivity already in the raw data. Figure 4.16b compares $\varepsilon_1(\omega)$ for BKFA at 12 K (with the itinerant response subtracted [39]) and RFS at 12 and 300 K (raw data). It is clear that the low-energy permittivity of RFS is about three times smaller than that of BKFA. This trend is very well reproduced in our LDA calculations shown in Fig. 4.16d, in which a comparable decrease is observed.

Our LDA calculations show, consistent with previous work [90, 91], that the stoichiometric 245 compounds are semiconducting and minor doping of either sign results in a very complicated Fermi surface in the magnetic state with only one type of carriers present. This is fully consistent with the weak metallic response of a nearly stoichiometric RFS observed in this work as well as with the overall phase diagram reported for this system in Ref. [85], where a narrow doping range was found for the superconducting phase bounded by an insulating and a semiconducting phase on the underdoped and overdoped sides, respectively. In the same Ref. [85] it was shown that the electronic specific-heat exhibits a rather small superconductivity-induced anomaly at the superconducting transition temperature. Together with the small effect of superconductivity on the itinerant optical response observed here, it implies that superconductivity in RFS is not a uniform bulk phenomenon. This conclusion is also consistent with a practically doping-independent superconducting transition temperature observed in Ref. [85] assuming that the superconducting phase stabilizes at the same doping level, while the excess is doped into the coexisting phase(s).

4.3.6 Conclusions

In summary, we obtained the complex dielectric function of a $Rb_2Fe_4Se_5$ superconductor with $T_c \approx 32$ K in the spectral range from 1 meV to 6.5 eV. Comparison with our LDA calculations shows that the optical response of this material is well reproduced and is close to its Fe–As based counterparts in the 122 family. Strikingly, unlike in the iron pnictides, the absorption band at 0.6 eV experiences spin-controlled narrowing into three sub-bands in the magnetic state. We further demonstrated that the superconducting RFS displays a clear metallic response in the THz spectral range below 100 K with $\omega_{pl} \approx 100$ meV, which can be divided into a narrow and a broad component and is partially suppressed in the superconducting state giving rise to a superconducting condensate with a plasma frequency of $\omega_{pl}^{SC} \approx 10$ meV. Such a small charge-carrier density suggests that the optical conductivity of the superconducting RFS represents an effective-medium response of two separate phases dominated by the magnetic semiconducting phase.

4.4 Nanoscale Phase Separation in $Rb_2Fe_4Se_5$

4.4.1 Introduction

The recent discovery of the intercalated iron-selenide superconductors[5] [73–76, 93] has stirred up the condensed-matter community accustomed to the proximity of the superconducting and magnetic phases in various cuprate and pnictide superconductors. Never before has a superconducting state with a transition temperature as high as 30 K been found to coexist with such an exceptionally strong antiferromagnetism with Néel temperatures up to 550 K as in this new family of iron-selenide materials. The very large magnetic moment of 3.3 μ_B on the iron sites [78], however, renders microscopically homogeneous coexistence of superconductivity and magnetism unlikely. Indeed, significant experimental evidence suggests that the superconducting and antiferromagnetic phases are spatially separated [71, 80–83, 86, 94].

In the prototypical iron arsenide superconductors $Ba(K)Fe(Co)_2As_2$ both the phase separation [95, 96] and coexistence [97] of antiferromagnetism and superconductivity have been shown to occur in certain regions of the phase diagram. At the same time, the structurally similar intercalated iron-selenide compounds $(K, Rb, Cs)_{0.8}Fe_{1.6}Se_2$ have defied all efforts to synthesize a bulk single-phase material of this family. Absence of such electronically homogeneous superconducting single crystals and a strong correlation between the superconducting and antiferromagnetic phases [87, 98] necessitate a detailed research into the nature of their coexistence. The volume fraction of the magnetic phase has been estimated to 86–88% in recent Mössbauer [86], bulk μSR [99], and NMR [100] studies. The shape of the phase domains, on the contrary, has seen much conflicting evidence with indications ranging from needlelike rather regular stripes [81] to insulating islands on a superconducting surface [94] to nanoseparated vacancy-disordered presumably metallic sheets in the bulk with an unknown in-plane form factor [101]. In a recent STM/STS study of [110] $K_xFe_{2-y}Se_2$ (KFS) thin films the superconducting phase was assigned

[5] The discussion in this section is largely based on Ref. [92].

to stoichiometric KFe_2Se_2 without iron vacancies [82]. However, consistent theoretical description of the recent INS and ARPES measurements based on this assumption requires significantly different levels of the chemical potential in the bulk and at the surface (equivalent to a disparity of ~0.08 electrons/Fe in the electron doping) [102, 103].

Infrared spectroscopy is well-suited to clearly distinguish between the antiferromagnetic semiconducting and paramagnetic metallic phases of RFS due to the large contrast in its complex dielectric function in this spectral range [71]. To determine the geometry of the domains and unambiguously assign them to the semiconducting and metallic phases, a submicrometer-resolution technique must be used. Here we employed s-SNOM in the infrared [104–106], which enables determination of the material's complex dielectric function with an unsurpassed in-plane resolution of ca 20 nm and a typical nanometer topographic sensitivity of AFM. We complemented this spatially-resolved technique with LE-μSR [107, 108] measurements on the same RFS single crystals to quantify the fraction of the magnetically-ordered phase in the bulk and trace its modification towards the sample surface.

4.4.2 Near-Field Optical Microscopy

The commercial s-SNOM near-field microscope uses an illuminated AFM probing tip to scan the sample surface and pseudo-heterodyne interferometric detection [109] to extract the near-field amplitude and phase from the light scattered back from the tip. Near-field spectroscopic experiments were performed with a commercial interferometric s-SNOM (NeaSNOM, www.neaspec.com), working in an intermittent contact mode. A standard platinum-coated AFM tip (NanoWorld ARROW-NCPt with a 25 nm radius) oscillating at a frequency $\Omega \approx 265\,kHz$ with an amplitude $\approx$25 nm, was illuminated with a focused infrared beam from a CO_2 laser (http://www.accesslaser.com) tuned to 10.7 μm and attenuated to approximately 5 mW. The light scattered back from the probing tip was detected with a HgCdTe detector (KLD-0,1-J1/DC/12, http://www.olmartech.com) in a pseudo-heterodyne interferometric mode, which enables simultaneous acquisition of both the amplitude and phase information. Signal demodulation at 2Ω with respect to the tapping frequency was used to extract the background-free near-field component from the detector signal.

The measurements were carried out on thin platelike (~5 × 5 × 0.5 mm^3) single crystals of optimally-doped superconducting RFS (batch BR26 in Ref. [85], $T_c \approx$ 32 K), cleaved prior to every scan. The sample surface was first characterized with a polarizing microscope. Figure 4.17a shows a 60 × 60 μm^2 patch of the sample surface. A network of bright stripes is always observed on a freshly-cleaved surface and does not depend on the polarization of the probing light. The stripes always occur at 45° with respect to the in-plane crystallographic axes. The same pattern was observed previously in Ref. [110] via a backscattered-electron analysis consistently on freshly-cleaved KFS samples. The topography of a representative 15 × 8 μm^2

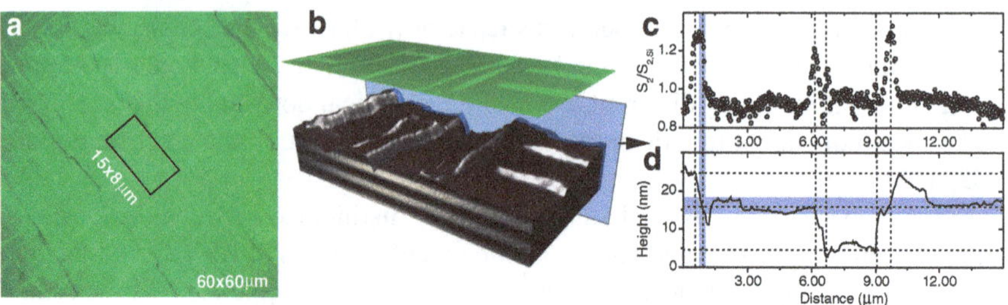

Fig. 4.17 **a** Microscope image of a $60 \times 60\,\mu m^2$ surface patch of a freshly-cleaved superconducting RFS single crystal. Typical rectangular $15 \times 8\,\mu m^2$ area studied via s-SNOM. **b** Superposition of the topography of a $15 \times 8\,\mu m^2$ rectangular area (terrain) and the optical signal (brightness) normalized to that of silicon. *Glossy areas* indicate high silicon-RFS contrast and thus metallicity, while the matt areas are insulating. This combined response is broken down in **c** and **d** for the cross-section defined by the blue semi-transparent plane. **c** Optical contrast $S_2/S_{2,\mathrm{Si}}$ of the second harmonic S_2 of the near-field signal obtained at the 10.7 μm emission wavelength (116 meV photon energy) of a CO_2 laser. Peaks in the contrast indicate metallic response (see text). **d** Displacement of the AFM tip while scanning along a 15 μm line obtained simultaneously with **c**. *Dashed lines* and *blue shaded areas* in **c** and **d** show the correlation between the metallic response and changes in the topography. Reprinted figure with permission from Ref. [92]. Copyright (2012) by the American Physical Society

surface patch studied with an s-SNOM microscope is shown in Fig. 4.17b as a three-dimensional terrain image. One can clearly correlate the features in the AFM map with the bright stripes in the polarizing microscope image. This implies that the surface chemistry of this compound leads to an inherent surface termination with mesoscopic terracing upon cleaving, typically 10–30 nm high. Figures 4.17c, d show the 2nd-harmonic near-field optical contrast (OC), obtained by normalizing the signal from the RFS surface to that that from a reference silicon surface ($S_2/S_{2,\mathrm{Si}}$), and the topography profile of RFS for the cross-section indicated with a blue translucent plane in Fig. 4.17b. The amplitude and phase of both the topography and the optical signal are obtained simultaneously during each scan. Every peak in the OC maps signals a metallic optical response. The absolute values of the complex dielectric function were obtained within the extended finite-dipole model [111] using OC maps. It provides an analytical expression for the near-field amplitude and phase, which can be used to calculate the detector signal at an arbitrary harmonic of the tip oscillation frequency. The model is based on the input parameters directly obtained in the experiment such as the curvature radius of the tip apex and the vibration frequency of the tip, which is assumed to be perfectly conductive. Since the model only addresses the relative contrast between different materials, the experimentally-obtained absolute near-field signals have to be normalized to those at the surface of a reference material. In our case all OC maps obtained on the RFS single crystals were normalized to those of bulk silicon measured in the same experimental geometry. The resulting relative optical amplitude $S_2/S_{2,\mathrm{Si}}$ and phase difference $\varphi_2 - \varphi_{2,\mathrm{Si}}$ were then analyzed in the finite-dipole model. The dielectric response of the semiconducting phase (dark regions in Fig. 4.17b and low OC in Fig. 4.17c) obtained in such a treatment of the experimental data ($\varepsilon_1 \approx 10$, $\varepsilon_2 \approx 0$) is fully consistent with that of the single-phase semiconducting RFS crystals [71]. The bright regions of the sample surface display negative values of ε_1, which provides solid evidence for their metallic character.

By correlating the OC peak positions with the topography maps one can deduce that the location of the metallic regions on the sample surface is bound to the slopes in the terrain, as indicated in Fig. 4.17c, d with dashed lines. Flat regions of the sample surface exhibit no metallic response. Figure 4.17d shows that from the location of the OC peaks on the slopes one can identify well-defined sheets (horizontal dashed lines) approximately 10 nm apart. Metallic response (OC peak) is recorded whenever one of these sheets is exposed from the sample bulk at the intersections of the dashed lines with the topography line. This correlation exists in all studied cross-sections of the OC and height maps. The thickness of the metallic sheets can be estimated by projecting the full width at half maximum (FWHM) of the OC peaks onto the surface topography (blue shaded areas in Fig. 4.17c, d) and on average amounts to 5 nm. The analysis of the correlations between the OC peaks and the slopes in the sample topography carried out on multiple cross-sections of different near-field maps shows that the metallic sheets have an in-plane dimension as well, which can be roughly estimated to $\approx 10\,\mu m$ in both directions. Therefore, phase separation in superconducting RFS single crystals occurs on the nanoscale in the out-of-plane and mesoscale in the in-plane direction. From the thickness and periodicity of the metallic sheets we can estimate the volume fraction of the metallic domains in the 30 nm surface layer to approximately 50 %, significantly larger than all previously reported bulk values.

All of the 245 iron-selenide superconductors known to date use elements of the first column of the periodic table as intercalants, which gives rise to high air reactivity of these compounds. The effects of surface oxidation have been observed with multiple techniques and carefully controlled in our measurements. Sample cleaving and preparation has been carried out in argon atmosphere at all times prior to every optical and LE-μSR measurement. In the latter case the sample matrix was loaded in the cryostat in argon atmosphere so that the sample surface was never exposed to air. In the case of near-field measurements it was impossible to avoid contact with air. Therefore, the time scale and effect of oxidation on the results of the measurements were studied and then monitored during the experiment. We found that first scans on a freshly-cleaved sample surface do not show any trace of ageing and noticeable oxidation typically occurs on the scale of hours. Figure 4.18 shows one of the maps in which significant oxidation was observed (the scan was taken nine hours after cleaving the surface). The surface map shown in the figure was obtained by scanning the surface from left to right, from top to bottom. The darker region in the left half of the image shows surface oxidation spreading towards the top right corner. The oxide film clearly affects the near-field signal, however, without completely eliminating the optical contrast between the metallic and semiconducting regions. Oxidation of the surface is expected to give rise to additional repulsive force exerted on the AFM tip and this effect has indeed been identified in the measurement. Close to the end of the scan the repulsive forces became so large that the set point of the AFM feedback loop had to be reduced. The effect of this change is clearly visible in the bottom of

Fig. 4.18 Near-field optical signal obtained on a $15 \times 8\ \mu m^2$ surface patch nine hours after cleaving. *Grainy darker region* of the image comes from the oxidized fraction of the surface and has lower optical signal and higher repulsive forces exerted on the AFM tip. Close to the end of the scan (bottom 15 % of the image) the set point of the AFM feedback loop had to be changed to overcome increasing repulsive forces from the oxidized surface

Fig. 4.18. In our final measurements the sample was cleaved prior to every scan and no surface ageing was observed within the accuracy of the instrument.

4.4.3 Low-Energy Muon-Spin Rotation/Relaxation

It must be noted that the metallic volume fraction obtained by means of s-SNOM imaging only considers the near surface layer within ≈30 nm. Whether or not this fraction changes with the distance from the sample surface and what the magnetic and superconducting properties of the metallic and semiconducting phases are cannot be determined from these measurements. Therefore, a microscopic sensor of the local magnetic moment with an adjustable implantation depth must be invoked. Such features are provided by the LE-μSR technique, which utilizes the predominantly spin-oriented positron decay of μ^+ to detect the orientation of the muon spin. This orientation is influenced by the local magnetic field at the muon stopping site, as well as by the magnetic fields applied externally. The experiment is shown schematically in Fig. 4.19a. Based on the results of the near-field measurements the sample is sketched as an idealized periodic layered structure, the finite in-plane dimension of the paramagnetic domains and the irregularity of their periodicity having been disregarded at this point.

In the LE-μSR technique, in contrast to its conventional counterpart, the incident energetic (~MeV) muon beam is first moderated to about 15 eV in a condensed layer of solid N_2 or Ar and then accelerated by a controlled electric field to achieve

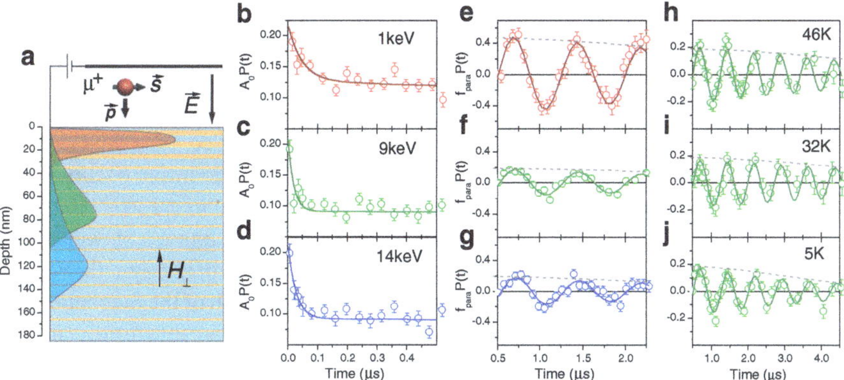

Fig. 4.19 **a** Schematic of the experiment with low-energy muons. The sample is depicted as a layered structure according to the discussion in the text. **b–d** Time dependence of μ^+ spin polarization $A_0P(t)$ in a zero magnetic field at 5 K after implantation in RFS single crystals at various depths determined by the muon stopping profiles of respective colors in **a**. **e–g** Same for normalized μ^+ spin polarization $f_{\text{para}}P(t)$ in a transverse magnetic field $H_\perp = 100$ G, where the normalized asymmetry f_{para} is the paramagnetic volume fraction. **h–j** Long-time dependence of normalized μ^+ spin polarization $f_{\text{para}}P(t)$ in a transverse magnetic field $H_\perp = 100$ G for the green muon stopping profile in **a** in the normal state at 46 K (**h**) and 32 K (**i**) and superconducting state at 5 K (**j**). *Gray dashed lines* in **e–j** show the slow relaxation envelope of the μ^+ spin polarization; *solid lines* are fits to the data. Reprinted figure with permission from Ref. [92]. Copyright (2012) by the American Physical Society

an adjustable implantation depth according to the muon stopping profile in a given material, calculated using the Monte Carlo algorithm TRIM.SP [112]. To study the dependence of the phase separation in RFS on the distance from the sample surface, we prepared a $\sim 15 \times 15\,\text{mm}^2$ matrix of single crystals and chose three muon stopping profiles shown as a red (achieved by accelerating the moderated muon beam to the energy of 1 keV), green (9 keV), and blue (14 keV) shaded profile in Fig. 4.19a. The resulting time dependence of μ^+ spin polarization $A_0P(t)$ at zero external magnetic field is shown in Fig. 4.19b–d, the colors correspond to those of the muon stopping profiles in Fig. 4.19a. The time dependence of the μ^+ spin polarization $A_0P(t)$ in zero external magnetic field was analyzed using a two-component depolarization function

$$A_0P(t) = A_{\text{s}}\exp(-\lambda_{\text{s}}t) + A_{\text{f}}\exp(-\lambda_{\text{f}}t),$$

where $A_{\text{s,f}}$ and $\lambda_{\text{s,f}}$ are the asymmetries and depolarization rates of the slow and fast components, respectively. In a fully isotropic magnetic environment $A_{\text{f}} = 2/3A_0$ and $A_{\text{s}} = 1/3A_0$. One third of all muons experience an internal field parallel to the spin so that the relaxation is only caused by dynamical fluctuations. If the fluctuation rate of the magnetic moments is small on the scale of the muon lifetime $\tau_{\mu^+} = 2.2\,\mu\text{s}$ then λ_{s} is small as well, as is the case for superconducting RFS single crystals. Two thirds of the muons feel an internal magnetic field transverse to their spin. In a para- or diamagnetic material this fast component does not exist.

Figure 4.19a shows that the shallowest muon stopping profile (red shaded area) is peaked at about 10 nm and extends to about 30 nm below the sample surface. It thus probes a region of the sample bulk comparable to that observed via s-SNOM. The relaxation of the muon polarization after implantation into the sample occurs at two different time scales: fast depolarization takes place during the first 150–250 ns and is associated with the ordered antiferromagnetic phase, while a much slower evolution dominates thereafter and stems predominantly from the paramagnetic volume fraction. The fast depolarization rate is proportional to the width of the magnetic-field distribution at the muon stopping site in the ordered phase and thus to the antiferromagnetic moment. Figure 4.19c, d show that in the bulk (green and blue muon stopping profiles in Fig. 4.19a) the fast component is approximately the same and rather narrow, whereas near the surface it broadens significantly (Fig. 4.19b). Using the aforementioned two-component fit model for the zero-field μSR data, the fast depolarization rates in the bulk were found to agree within the error bars: 52(13) µs and 37(7) µs for the green and blue profiles, respectively, whereas it is significantly reduced to 21(4) µs close to the surface (red profile). Under the assumption that the muon stopping sites in the unit cell are the same for all three profiles and that the depolarization rate should be the same in the bulk, the antiferromagnetic moment is reduced to only 50 % of its bulk value in the 30 nm surface layer. This reduction might originate in electronic-density redistribution due to dangling bonds and a possible surface segregation of ion vacancies.

The muon decay asymmetry in Fig. 4.19b–d indicates that the paramagnetic volume fraction (slow component) is enhanced closer to the surface. To avoid the contaminating μSR signal originating from muon decays in the Ni sample holder, we carried out transverse-field measurements in the same configuration and at the same temperature of 5 K as in Fig. 4.19b–d. The time dependence of the μ^+ spin polarization $A_0 P(t)$ in an external transverse magnetic field was analyzed after discarding the fast muon relaxation in the ordered antiferromagnetic phase during the first 150 ns. The remaining slow depolarization characteristic of a paramagnetic state was fitted with a single Gaussian function after factoring in an oscillatory term to account for the precession of the muon spin in the external magnetic field (see Sect. 3.3):

$$A_{\mathrm{TF}} P(t) = A_{\mathrm{TF}} \exp\left[-\frac{1}{2}\left(\sigma_{\mathrm{sc}}^2 + \sigma_{\mathrm{n}}^2\right) t^2\right] \cos(\gamma_\mu B_{\mathrm{int}}\, t + \varphi),$$

where A_{TF} is the initial asymmetry, B_{int} is the internal magnetic field at the muon site (close to the external applied magnetic field), $\sigma_{\mathrm{n}}, \sigma_{\mathrm{sc}}$ are the Gaussian depolarization rates due to the nuclear magnetic moments and due to the magnetic field distribution in the vortex state, respectively. The depolarization rate σ_{n} was extracted from the normal-state data right above the superconducting transition temperature and considered temperature-independent down to 5 K; φ is a constant detector phase.

In the case of transverse-field measurements the oscillating component of the muon decay asymmetry comes only from the paramagnetic domains in the sample and the paramagnetic volume fraction f_{para} can be extracted from the fitted asymmetry A_{TF} using the following relation:

$$f_{para} = \frac{A_{TF}}{S \times A_{TF}^{max}(E)},$$

where S is the fraction of the beam covered by the sample, and $A_{TF}^{max}(E)$ is the instrumental energy-dependent maximum asymmetry in a transverse field equal for our experimental setup to 0.155, 0.245, and 0.250 at 1, 9, and 14 keV, respectively. The oscillating component of the muon decay asymmetry can thus be normalized accordingly to become $f_{para}P(t)$, where f_{para} is the paramagnetic volume fraction, equal to the value of the slowly depolarizing envelope of the muon spin precession (gray dashed lines in Fig. 4.19) at zero time.

From Fig. 4.19f, g one immediately infers that well in the sample bulk the paramagnetic phase constitutes about 20 % of the sample volume. It thus characterizes the phase separation in the bulk of the superconducting RFS single crystals as probed by other techniques [86, 99, 100]. On the other hand, Fig. 4.19e shows that in the 30 nm surface layer (red stopping profile in Fig. 4.19a) the paramagnetic volume fraction strongly increases to ≈50 %. Such a high value most likely occurs due to a reduced constraining potential between antiferromagnetic sheets with a significantly smaller ordered moment close to the sample surface, which leads to an expansion of the metallic regions. These depth-dependent physical properties could explain the difference in the chemical potential required for a consistent interpretation of the data obtained with bulk (INS) and surface-sensitive (ARPES) probes.

To characterize the superconducting phase of RFS with μSR we studied the temperature dependence of the oscillating component of the muon asymmetry in a vortex state generated by the transverse external magnetic field, perpendicular to the metallic sheets. The oscillations are damped by the inhomogeneity of the magnetic field in the vortices, which, in turn, is proportional to the condensate density $n_s \propto 1/\lambda_{ab}^2$. The measurements were carried out in a transverse magnetic field $H_\perp = 100$ G for the muon stopping profile peaked at 80 nm below the sample surface (green shaded area in Fig. 4.19a). The time dependence of the normalized μ^+ spin polarization $f_{para}P(t)$ in the normal state at 46 and 32 K and in the superconducting state at 5 K is shown in Fig. 4.19h–j, respectively. It is evident from the data that the damping in the normal state is approximately constant, while in the superconducting state it is noticeably faster, which indicates the presence of a superconducting condensate. By subtracting the normal-state damping from that in the superconducting state we can estimate the London penetration depth $\lambda_{||} \approx 550$ nm. Assuming that the phase separation in the bulk has the same layered structure as detected near the surface, this value must be revisited in an appropriate model. For a stack of superconducting layers of thickness d_{sc} separated by insulating layers of thickness d_{ins} the inhomogeneity of the magnetic field in a vortex (which is then a stack of two-dimensional vortex pancakes) is described by the Lawrence-Doniach in-plane penetration depth $\lambda_{||}$, related to the bulk London penetration depth λ_{ab} via $\lambda_{||} = \lambda_{ab}(d_{sc}+d_{ins})^{1/2}/d_{sc}^{1/2}$ (see Ref. [113]).

This reduction reflects the fact that $\lambda_{||}$ is related to the *average* superconducting condensate density $\langle n_s \rangle = n_s d_{sc}/(d_{sc} + d_{ins}) = n_s f_{para}$. Taking these considerations into account and using the bulk paramagnetic volume fraction $f_{para} = 0.2$ obtained in our LE-μSR measurements one can estimate the intrinsic bulk in-plane London penetration depth of the superconducting phase to be $\lambda_{ab} \approx 250\,\text{nm}$. This value would increase towards $\lambda_{||}$ due to the finite in-plane dimension of the superconducting layers and their disordered stacking since the superconducting phase inclusions would then contribute independently to the μSR depolarization. At the same time, our procedure slightly overestimates the value of $\lambda_{||}$ due to the widening of the vortex field distribution close to the surface [114].

Similar effective-medium approximations (EMA) must be used for an adequate analysis of the results obtained with other experimental techniques. A recent study of the optical conductivity of the same superconducting single crystals [71] emphasized the importance of EMA but could not make a concrete estimate due to the unknown details of the phase separation in this compound. It reported the total plasma frequency of the itinerant charge carriers of about 100 meV. Using the same dimensions for the metallic paramagnetic and semiconducting antiferromagnetic layers as in the interpretation of the LE-μSR results one can extract the bulk optical response from the conductivity data reported in Ref. [71]. A simple fit of the experimental data as the optical response of a perfect superlattice gives a significantly larger value of the total plasma frequency $\omega_{pl}^{tot} \approx 300\,\text{meV}$ and brings the estimated value of the London penetration depth $\lambda_{ab}^{opt} \approx 2\,\mu\text{m}$ closer to that obtained with LE-μSR but still 8 times larger. Accounting for the finite in-plane dimension of the paramagnetic domains is expected to lead to a better agreement between the two techniques.

4.4.4 Conclusions

The origin of the phase separation observed in this work can lie either in chemical stratification into e.g. iron-vacancy ordered (antiferromagnetic) and disordered (paramagnetic) phases or in purely electronic segregation on a homogeneous crystalline background. Self-organization of a chemically homogeneous structure into (quasi)periodically segregated phases is not unprecedented—a similar phenomenon has been observed in copper-based superconductors, where antiferromagnetic stripes of copper spins were found to be spatially separated by periodic domain walls close to a particular hole doping level of 1/8 [115], albeit with a quite different characteristic length scale. In the case of chemical inhomogeneity on a nanometer scale out of plane, the phases are likely to exert a significant internal pressure on each other, thus changing the c-axis lattice parameter and, consequently, the Fe-Se bond angle. The latter is known to have a strong effect on the superconducting and magnetic properties of the iron-based superconductors. Be it of chemical or electronic nature, the phase separation in superconducting RFS single crystals reported here represents an interesting case of a naturally-occurring quasi-heterostructure. It is also noteworthy, that the out-of-plane dimension of the superconducting domains is comparable to the c-axis coherence length $\xi_c^{RFS} \approx 1$ nm, as determined from the upper critical fields $H_{c2}^{ab,c} = 63$ and 25 T, respectively, obtained in Ref. [85].

To conclude, by combining the unique optical imaging capabilities and nanoscale resolution of the s-SNOM near-field microscope with the bulk sensitivity at variable depth of LE-μSR we determined the geometry and magnitude of phase separation in RFS superconducting single crystals. The paramagnetic domains were found to have the shape of thin metallic sheets parallel to the iron-selenide planes of the crystal with a characteristic size of only several nanometers out-of-plane but up to 10 μm in-plane. By means of LE-μSR we further showed that the antiferromagnetic semiconducting phase occupies ≈80 % of the sample volume in the bulk and is strongly weakened near the surface. These results have important implications for the interpretation of bulk- and surface-sensitive measurements on $Rb_2Fe_4Se_5$, and for the understanding of the interplay between superconductivity and antiferromagnetism in this material.

References

1. Charnukha, A., et al. (2011). Superconductivity-induced optical anomaly in an iron arsenide. *Nature Communication, 2*, 219.
2. Allender, D., Bray, J., & Bardeen, J. (1973). Model for an exciton mechanism of superconductivity. *Physical Review B, 7*, 1020–1029.
3. Little, W. A. (1964). Possibility of synthesizing an organic superconductor. *Physical Review, 134*, A1416–A1424.
4. Ginzburg, V. L. (1965). Concerning surface superconductivity. *Soviet Physics JETP, 20*, 1549.
5. Littlewood, P. B., et al. (2004). Models of coherent exciton condensation. *Journal of Physics: Condensed Matter, 16*, S3597–S3620.
6. Hirsch, J. E., & Marsiglio, F. (1989). Superconducting state in an oxygen hole metal. *Physical Review B, 39*, 11515–11525.
7. Basov, D. N., & Timusk, T. (2005). Electrodynamics of high-T_c superconductors. *Reviews of Modern Physics, 77*, 721–779.
8. Maiti, S., & Chubukov, A. V. (2010). Optical integral and sum-rule violation in high-T_c superconductors. *Physical Review B, 81*, 245111.
9. Holcomb, M. J., Perry, C. L., Collman, J. P., & Little, W. A. (1996). Thermal-difference reflectance spectroscopy of the high-temperature cuprate superconductors. *Physical Review B, 53*, 6734–6751.
10. Boris, A. V., et al. (2004). In-plane spectral weight shift of charge carriers in $YBa_2Cu_3O_{6.9}$. *Science, 304*, 708.
11. Kuzmenko, A. B., Molegraaf, H. J. A., Carbone, F., & van der Marel, D. (2005). Temperature-modulation analysis of superconductivity-induced transfer of in-plane spectral weight in $Bi_2Sr_2CaCu_2O_8$. *Physical Review B, 72*, 144503.
12. Toschi, A., et al. (2005). Temperature dependence of the optical spectral weight in the cuprates: Role of electron correlations. *Physical Review Letters, 95*, 097002.
13. Haule, K., & Kotliar, G. (2007). Strongly correlated superconductivity: A plaquette dynamical mean-field theory study. *Physical Review B, 76*, 104509.
14. Mazin, I. I. (2010). Superconductivity gets an iron boost. *Nature, 464*, 183.
15. Boeri, L., Dolgov, O. V., & Golubov, A. A. (2008). Is $LaFeAsO_{1-x}F_x$ an electron-phonon superconductor? *Physical Review Letters, 101*, 026403.

16. Mazin, I. I., Singh, D. J., Johannes, M. D., & Du, M. H. (2008). Unconventional superconductivity with a sign reversal in the order parameter of $LaFeAsO_{1-x}F_x$. *Physical Review Letters, 101*, 057003.
17. Popovich, P., et al. (2010). Specific heat measurements of $Ba_{0.68}K_{0.32}Fe_2As_2$ single crystals: evidence for a multiband strong-coupling superconducting state. *Physical Review Letters, 105*, 027003.
18. Li, G., et al. (2008). Probing the superconducting energy gap from infrared spectroscopy on a $Ba_{0.6}K_{0.4}Fe_2As_2$ single crystal with $T_c = 37$ K. *Physical Review Letters, 101*, 107004.
19. Maldague, P. F. (1977). Optical spectrum of a Hubbard chain. *Physical Review B, 16*, 2437–2446.
20. Hirsch, J. (1992). Apparent violation of the conductivity sum rule in certain superconductors. *Physica C, 199*, 305–310.
21. Andersen, O. K. (1975). Linear methods in band theory. *Physical Review B, 12*, 3060–3083.
22. Rotter, M., et al. (2008). Spin-density-wave anomaly at 140 K in the ternary iron arsenide $BaFe_2As_2$. *Physical Review B, 78*, 020503.
23. Tegel, M. et al. (2008). Structural and magnetic phase transitions in the ternary iron arsenides $SrFe_2As_2$ and $EuFe_2As_2$. *Journal of Physics: Condensed Matter, 20*, 452201.
24. Analytis, J. G., Chu, J.-H., McDonald, R. D., Riggs, S. C., & Fisher, I. R. (2010). Enhanced Fermi-surface nesting in superconducting $BaFe_2(As_{1-x}P_x)_2$ revealed by de Haas-van Alphen effect. *Physical Review Letters, 105*, 207004.
25. Analytis, J. G., et al. (2009). Fermi surface of $SrFe_2P_2$ determined by the de Haas-van Alphen effect. *Physical Review Letters, 103*, 076401.
26. Yi, M., et al. (2009). Electronic structure of the $BaFe_2As_2$ family of iron-pnictide superconductors. *Physical Review B, 80*, 024515.
27. Skornyakov, S. L., et al. (2009). Classification of the electronic correlation strength in the iron pnictides: The case of the parent compound $BaFe_2As_2$. *Physical Review B, 80*, 092501.
28. Hu, W. Z., et al. (2008). Origin of the spin density wave instability in $AFe_2As_2 (A = \mathrm{Ba, Sr})$ as revealed by optical spectroscopy. *Physical Review Letters, 101*, 257005.
29. Shimojima, T., et al. (2010). Orbital-dependent modifications of electronic structure across the magnetostructural transition in $BaFe_2As_2$. *Physical Review Letters, 104*, 057002.
30. Tinkham, M. (1995). *Introduction to superconductivity*. New York: McGraw-Hill.
31. Dobryakov, A. L., Farztdinov, V. M., Lozovik, Y. E., & Letokhov, V. S. (1994). Energy gap in the superconductor optical spectrum. *Optics Communication, 105*, 309–314.
32. Suhl, H., Matthias, B. T., & Walker, L. R. (1959). Bardeen-Cooper-Schrieffer theory of superconductivity in the case of overlapping bands. *Physical Review Letters, 3*, 552–554.
33. Berciu, M., Elfimov, I., & Sawatzky, G. A. (2009). Electronic polarons and bipolarons in iron-based superconductors: The role of anions. *Physical Review B, 79*, 214507.
34. Sawatzky, G. A., Elfimov, I. S., van den Brink, J., & Zaanen, J. (2009). Heavy-anion solvation of polarity fluctuations in pnictides. *Europhysics Letters, 86*, 17006.
35. Bardeen, J., Cooper, L. N., & Schrieffer, J. R. (1957). Theory of superconductivity. *Physical Review, 108*, 1175–1204.
36. Evtushinsky, D. V., et al. (2009). Momentum dependence of the superconducting gap in $Ba_{1-x}K_xFe_2As_2$. *Physical Review B, 79*, 054517.
37. Lee, W. S., et al. (2007). Abrupt onset of a second energy gap at the superconducting transition of underdoped Bi2212. *Nature, 450*, 81–84.
38. Campuzano, J. C., et al. (1996). Direct observation of particle-hole mixing in the superconducting state by angle-resolved photoemission. *Physical Review B, 53*, R14737.
39. Charnukha, A., et al. (2011). Eliashberg approach to infrared anomalies induced by the superconducting state of $Ba_{0.68}K_{0.32}Fe_2As_2$ single crystals. *Physical Review B, 84*, 174511.
40. Kamihara, Y., Watanabe, T., Hirano, M., & Hosono, H. (2008). Iron-based layered superconductor $La[O_{1-x}F_x]FeAs (x = 0.05 - 0.12)$ with $T_c = 26$K. *Journal of the American Chemical Society, 130*, 3296.
41. Paglione, J., & Greene, R. L. (2010). High-temperature superconductivity in iron-based materials. *Nature Physics, 6*, 645.

42. Johnston, D. C. (2010). The puzzle of high temperature superconductivity in layered iron pnictides and chalcogenides. *Advances in Physics, 59*, 803.
43. Zhang, Y., et al. (2010). Out-of-plane momentum and symmetry-dependent energy gap of the pnictide $Ba_{0.6}K_{0.4}Fe_2As_2$ superconductor revealed by angle-resolved photoemission spectroscopy. *Physical Review Letters, 105*, 117003.
44. Nakayama, K., et al. (2011). Universality of superconducting gaps in overdoped $Ba_{0.3}K_{0.7}Fe_2As_2$ observed by angle-resolved photoemission spectroscopy. *Physical Review B, 83*, 020501.
45. Xu, Y.-M., et al. (2011).Observation of a ubiquitous three-dimensional superconducting gap function in optimally doped $Ba_{0.6}K_{0.4}Fe_2As_2$. *Nature Physics, 7*, 198.
46. Shan, L. et al. (2011). Observation of ordered vortices with Andreev bound states in $Ba_{0.6}K_{0.4}Fe_2As_2$. *Nature Physics 7*, 325.
47. Shan, L., et al. (2011). Evidence of multiple nodeless energy gaps in superconducting $Ba_{0.6}K_{0.4}Fe_2As_2$ single crystals from scanning tunneling spectroscopy. *Physical Review B, 83*, 060510.
48. Sun, G. L., et al. (2011). Single crystal growth and effect of doping on structural, transport and magnetic properties of $A_{1-x}K_xFe_2As_2$ (A =Ba, Sr). *Journal of Superconductivity and Novel Magnetism, 24*, 1773.
49. Boris, A. V., et al. (2009). Signatures of electronic correlations in optical properties of $LaFeAsO_{1-x}F_x$. *Physical Review Letters, 102*, 027001.
50. Tu, J. J., et al. (2010). Optical properties of the iron arsenic superconductor $BaFe_{1.85}Co_{0.15}As_2$. *Physical Review B, 82*, 174509.
51. Kim, K. W., et al. (2010). Evidence for multiple superconducting gaps in optimally doped $BaFe_{1.87}Co_{0.13}As_2$ from infrared spectroscopy. *Physical Review B, 81*, 214508.
52. Lobo, R. P. S. M., et al. (2010). Optical signature of subgap absorption in the superconducting state of $Ba(Fe_{1-x}Co_x)_2As_2$. *Physical Review B, 82*, 100506.
53. Zimmermann, W., Brandt, E., Bauer, M., Seider, E., & Genzel, L. (1991). Optical conductivity of BCS superconductors with arbitrary purity. *Physica C, 183*, 99.
54. Nam, S. B. (1967). Theory of electromagnetic properties of superconducting and normal systems. *I. Physical Review, 156*, 470.
55. Shulga, S., Dolgov, O., & Maksimov, E. (1991). Electronic states and optical spectra of HTSC with electron-phonon coupling. *Physica C, 178*, 266.
56. Armitage, N. P., et al. (2010). Infrared conductivity of elemental bismuth under pressure: Evidence for an avoided Lifshitz-type semimetal-semiconductor transition. *Physical Review Letters, 104*, 237401.
57. Nam, S. B. (1967). Theory of electromagnetic properties of strong-coupling and impure superconductors. *II. Physical Review, 156*, 487.
58. Dolgov, O. V., Kremer, R. K., Kortus, J., Golubov, A. A., & Shulga, S. V. (2005). Thermodynamics of two-band superconductors: The case of MgB_2. *Physical Review B, 72*, 024504.
59. Golubov, A. A., & Mazin, I. I. (1997). Effect of magnetic and nonmagnetic impurities on highly anisotropic superconductivity. *Physical Review B, 55*, 15146.
60. Chen, G. F., et al. (2008). Transport and anisotropy in single-crystalline $SrFe_2As_2$ and $A_{0.6}K_{0.4}Fe_2As_2$ (A = Sr, Ba) superconductors. *Physical Review B, 78*, 224512.
61. Rotter, M., Tegel, M., & Johrendt, D. (2008). Superconductivity at 38 K in the iron arsenide $(Ba_{1-x}K_x)Fe_2As_2$. *Physical Review Letters, 101*, 107006.
62. Mu, G., et al. (2009). Low temperature specific heat of the hole-doped $Ba_{0.6}K_{0.4}Fe_2As_2$ single crystals. *Physical Review B, 79*, 174501.
63. Christianson, A. D., et al. (2008). Unconventional superconductivity in $Ba_{0.6}K_{0.4}Fe_2As_2$ from inelastic neutron scattering. *Nature, 456*, 930.
64. Inosov, D. S., et al. (2009). Normal-state spin dynamics and temperature-dependent spin-resonance energy in optimally doped $BaFe_{1.85}Co_{0.15}As_2$. *Nature Physics, 6*, 178.
65. Nicol, E. J., Carbotte, J. P., & Timusk, T. (1991). Optical conductivity in high-T_c superconductors. *Physical Review B, 43*, 473–479.

66. Charnukha, A. (2011). Interactive simulation. http://www.fkf.mpg.de/keimer/groups/optical/bkfafir.jar.
67. Fang, L., et al. (2009). Roles of multiband effects and electron-hole asymmetry in the superconductivity and normal-state properties of $Ba(Fe_{1-x}Co_x)_2As_2$. *Physical Review B, 80*, 140508.
68. Coldea, A. I., et al. (2008). Fermi surface of superconducting LaFePO determined from quantum oscillations. *Physical Review Letters, 101*, 216402.
69. Tortello, M., et al. (2010). Multigap superconductivity and strong electron-boson coupling in Fe-based superconductors: A point-contact Andreev-reflection study of $Ba(Fe_{1-x}Co_x)_2As_2$ single crystals. *Physical Review Letters, 105*, 237002.
70. Hardy, F., et al. (2010). Doping evolution of superconducting gaps and electronic densities of states in $Ba(Fe_{1-x}Co_x)_2As_2$ iron pnictides. *Europhysics Letters, 91*, 47008.
71. Charnukha, A., et al. (2012). Optical conductivity of superconducting $Rb_2Fe_4Se_5$ single crystals. *Physical Review B, 85*, 100504.
72. Barišić, N., et al. (2010). Electrodynamics of electron-doped iron pnictide superconductors: Normal-state properties. *Physical Review B, 82*, 054518.
73. Guo, J., et al. (2010). Superconductivity in the iron selenide $K_xFe_2Se_2$ $(0 \leq x \leq 1.0)$. *Physical Review B, 82*, 180520.
74. Ying, J. J., et al. (2011). Superconductivity and magnetic properties of single crystals of $K_{0.75}Fe_{1.66}Se_2$ and $Cs_{0.81}Fe_{1.61}Se_2$. *Physical Review B, 83*, 212502.
75. Mizuguchi, Y., et al. (2011). Transport properties of the new Fe-based superconductor $K_xFe_2Se_2$ ($T_c = 33$K). *Applied Physics Letters, 98*, 042511.
76. Wang, A. F., et al. (2011). Superconductivity at 32 k in single-crystalline $Rb_xFe_{2-y}Se_2$. *Physical Review B, 83*, 060512.
77. Bacsa, J., et al. (2011). Cation vacancy order in the $K_{0.8+x}Fe_{1.6-y}Se_2$ system: Five-fold cell expansion accommodates 20% tetrahedral vacancies. *Chemical Sciences, 2*, 1054.
78. Bao, W., et al. (2011). A novel large moment antiferromagnetic order in $K_{0.8}Fe_{1.6}Se_2$ superconductor. *Chinese Physics Letters, 28*, 086104.
79. Park, J. T., et al. (2011). Magnetic resonant mode in the low-energy spin-excitation spectrum of superconducting $Rb_2Fe_4Se_5$ single crystals. *Physical Review Letters, 107*, 177005.
80. Ricci, A., et al. (2011). Nanoscale phase separation in the iron chalcogenide superconductor $K_{0.8}Fe_{1.6}Se_2$ as seen via scanning nanofocused X-ray diffraction. *Physical Review B, 84*, 060511.
81. Yuan, R. H., et al. (2012). Nanoscale phase separation of antiferromagnetic order and superconductivity in $K_{0.75}Fe_{1.75}Se_2$. *Science Report, 2*, 221.
82. Li, W., et al. (2012). Phase separation and magnetic order in K-doped iron selenide superconductor. *Nature Physics, 8*, 126–130.
83. Wang, M., et al. (2011). Antiferromagnetic order and superlattice structure in nonsuperconducting and superconducting $Rb_yFe_{1.6+x}Se_2$. *Physical Review B, 84*, 094504.
84. Chen, Z. G., et al. (2011). Infrared spectrum and its implications for the electronic structure of the semiconducting iron selenide $K_{0.83}Fe_{1.53}Se_2$. *Physical Review B, 83*, 220507.
85. Tsurkan, V., et al. (2011). Anisotropic magnetism, superconductivity, and the phase diagram of $Rb_{1-x}Fe_{2-y}Se_2$. *Physical Review B, 84*, 144520.
86. Ksenofontov, V., et al. (2011). Phase separation in superconducting and antiferromagnetic $Rb_{0.8}Fe_{1.6}Se_2$ probed by Mössbauer spectroscopy. *Physical Review B, 84*, 180508.
87. Ye, F., et al. (2011). Common crystalline and magnetic structure of superconducting $A_2Fe_4Se_5$ (A = K, Rb, Cs, Tl) single crystals measured using neutron diffraction. *Physical Review Letters, 107*, 137003.
88. Coldea, A. I., et al. (2009). Topological change of the Fermi surface in ternary iron pnictides with reduced c/a ratio: A de Haas-van Alphen study of $CaFe_2P_2$. *Physical Review Letters, 103*, 026404.
89. Shishido, H., et al. (2010). Evolution of the Fermi surface of $BaFe_2(As_{1-x}P_x)_2$ on entering the superconducting dome. *Physical Review Letters, 104*, 057008.

90. Yan, X.-W., Gao, M., Lu, Z.-Y., & Xiang, T. (2011). Ternary iron selenide $K_{0.8}Fe_{1.6}Se_2$ is an antiferromagnetic semiconductor. *Physical Review B, 83*, 233205.
91. Cao, C., & Dai, J. (2011). Block spin ground state and three-dimensionality of $(K, Tl)_yFe_{1.6}Se_2$. *Physical Review Letters, 107*, 056401.
92. Charnukha, A., et al. (2012). Nanoscale layering of antiferromagnetic and superconducting phases in $Rb_2Fe_4Se_5$ single crystals. *Physical Review Letters, 109*, 017003.
93. Zhang, Y., et al. (2011). Nodeless superconducting gap in $A_xFe_2Se_2$ (A = K, Cs) revealed by angle-resolved photoemission spectroscopy. *Nature Materials, 10*, 273–277.
94. Chen, F., et al. (2011). Electronic identification of the parental phases and mesoscopic phase separation of $K_xFe_{2-y}Se_2$ superconductors. *Physical Review X, 1*, 021020.
95. Park, J. T., et al. (2009). Electronic phase separation in the slightly underdoped iron pnictide superconductor $Ba_{1-x}K_xFe_2As_2$. *Physical Review Letters, 102*, 117006.
96. Inosov, D. S., et al. (2009). Suppression of the structural phase transition and lattice softening in slightly underdoped $Ba_{1-x}K_xFe_2As_2$ with electronic phase separation. *Physical Review B, 79*, 224503.
97. Marsik, P., et al. (2010). Coexistence and competition of magnetism and superconductivity on the nanometer scale in underdoped $BaFe_{1.89}Co_{0.11}As_2$. *Physical Review Letters, 105*, 057001.
98. Ksenofontov, V., et al. (2012). Superconductivity and magnetism in $Rb_{0.8}Fe_{1.6}Se_2$ under pressure. *Physical Review B, 85*, 214519.
99. Shermadini, Z., et al. (2012). Superconducting properties of single-crystalline $A_xFe_{2-y}Se_2$ (A = Rb,K) studied using muon spin spectroscopy. *Physical Review B, 85*, 100501.
100. Texier, Y., et al. (2012). NMR study in the iron-selenide $Rb_{0.74}Fe_{1.6}Se_2$: Determination of the superconducting phase as iron vacancy-free $Rb_{0.3}Fe_2Se_2$. *Physical Review Letters, 108*, 237002.
101. Wang, Z., et al. (2011). Microstructure and ordering of iron vacancies in the superconductor system $K_yFe_xSe_2$ as seen via transmission electron microscopy. *Physical Review B, 83*, 140505.
102. Friemel, G., et al. (2012). Reciprocal-space structure and dispersion of the magnetic resonant mode in the superconducting phase of $Rb_xFe_{2-y}Se_2$ single crystals. *Physical Review B, 85*, 140511.
103. Maier, T. A., Graser, S., Hirschfeld, P. J., & Scalapino, D. J. (2011). d-wave pairing from spin fluctuations in the $K_xFe_{2-y}Se_2$ superconductors. *Physical Review B, 83*, 100515.
104. Hillenbrand, R., Taubner, T., & Keilmann, F. (2002). Phonon-enhanced light-matter interaction at the nanometre scale. *Nature, 418*, 159–162.
105. Keilmann, F., & Hillenbrand, R. (2008). *Nano-optics and near-field optical microscopy*. London: Artech House.
106. Huth, F., Schnell, M., Wittborn, J., Ocelic, N., & Hillenbrand, R. (2011). Infrared-spectroscopic nanoimaging with a thermal source. *Nature Materials, 10*, 352–356.
107. Prokscha, T., et al. (2008). The new μE4 beam at PSI: A hybrid-type large acceptance channel for the generation of a high intensity surface-muon beam. *Nuclear Instruments and Methods in Physics Research Section A, 595*, 317–331.
108. Boris, A. V., et al. (2011). Dimensionality control of electronic phase transitions in nickel-oxide superlattices. *Science, 332*, 937–940.
109. Ocelic, N., Huber, A., & Hillenbrand, R. (2006). Pseudoheterodyne detection for background-free near-field spectroscopy. *Applied Physics Letters, 89*, 101124.
110. Ryan, D. H., et al. (2011). ^{57}Fe mössbauer study of magnetic ordering in superconducting $K_{0.80}Fe_{1.76}Se_{2.00}$ single crystals. *Physical Review B, 83*, 104526.
111. Cvitkovic, A., Ocelic, N., & Hillenbrand, R. (2007). Analytical model for quantitative prediction of material contrasts in scattering-type near-field optical microscopy. *Optics Express, 15*, 8550–8565.
112. Morenzoni, E., et al. (2002). Implantation studies of keV positive muons in thin metallic layers. *Nuclear Instruments and Methods in Physics Research B, 192*, 254–266.
113. Clem, J. R. (1991). Two-dimensional vortices in a stack of thin superconducting films: A model for high-temperature superconducting multilayers. *Physical Review B, 43*, 7837–7846.

114. Niedermayer, C., et al. (1999). Direct observation of a flux line lattice field distribution across an $YBa_2Cu_3O_{7-\delta}$ surface by low energy muons. *Physical Review Letters*, *83*, 3932.
115. Tranquada, J. M., Sternlieb, B. J., Axe, J. D., Nakamura, Y., & Uchida, S. (1995). Evidence for stripe correlations of spins and holes in copper oxide superconductors. *Nature*, *375*, 561–563.

Chapter 5
Summary

> *Everything should be made as simple as possible, but not one bit simpler.*
>
> —Albert Einstein

In this chapter the main experimental results and the new insight into the mechanism of superconductivity in the iron-based materials obtained in this work are briefly summarized. We studied the prototypical $ThCr_2Si_2$-type iron-arsenide superconductor $Ba_{0.68}K_{0.32}Fe_2As_2$ and its selenium-based counterpart, $Rb_2Fe_4Se_5$, by means of spectroscopic ellipsometry and, in the latter case, time-domain transmission spectroscopy, s-SNOM, and LE-μSR. Although these materials have a very similar crystallographic structure the former has been obtained in a single-crystalline form whereas the latter has defied all such efforts and even the best samples synthesized remain phase-separated.

One of the central tenets of conventional theories of superconductivity, including most models proposed for the recently discovered iron-pnictide superconductors, is the notion that only electronic excitations with energies comparable to the superconducting energy gap are affected by the transition. Here we reported the results of a comprehensive spectroscopic ellipsometry study of a high-quality crystal of superconducting $Ba_{0.68}K_{0.32}Fe_2As_2$ that challenge this notion. We observed a superconductivity-induced suppression of an absorption band at an energy of 2.5 eV, two orders of magnitude above the superconducting gap energy $2\Delta \sim 20$ meV. Based on density-functional calculations, this band could be assigned to transitions from As-p to Fe-d orbitals crossing the Fermi level. We identified a related effect at the spin-density–wave transition in parent compounds of the 122 family. We argued that the observed superconductivity-induced suppression involves a redistribution of electronic population between bands with different orbital character. Our results emphasize the importance of orbital physics for the mechanism of superconductivity in the iron pnictides. This conclusion is strongly supported by the recent ARPES observation of a clear correlation between the orbital character of the underlying sheets of the Fermi surface and the magnitude and dispersion

A. Charnukha, *Charge Dynamics in 122 Iron-Based Superconductors*,
Springer Theses, DOI: 10.1007/978-3-319-01192-9_5,

of the superconducting energy gap in $Ba_{0.6}K_{0.4}Fe_2As_2$ [1], orbital-selective band renormalization in $Ba(Fe_{1-x}Co_x)_2As_2$ [2], and orbital-dependent modification of the Fermi surface across the structural transition in $BaFe_2As_2$ [3] to name a few. Several proposals for orbital-fluctuation mediated superconductivity have been put forward [4, 5]. Our results also suggest that As-p states deep below the Fermi level might contribute to the formation of the superconducting and spin-density–wave states in the iron arsenides.

Based on our measurements of the full complex dielectric function of high-purity $Ba_{0.68}K_{0.32}Fe_2As_2$ single crystals with $T_c = 38.5$ K by broadband spectroscopic ellipsometry at temperatures $10 \leq T \leq 300$ K, we then discussed the microscopic origin of superconductivity-induced infrared optical anomalies in the framework of a multiband Eliashberg theory with two distinct superconducting energy gaps $2\Delta_A \approx 6\ k_B T_c$ and $2\Delta_B \approx 2.2\ k_B T_c$. This analysis allowed us to ascribe the observed unusual suppression of the optical conductivity in the superconducting state at energies up to 14 $k_B T_c$ to spin-fluctuation–assisted processes in the clean limit of the strong-coupling regime. Our results explain that the fact that no such anomaly has been observed in $Ba(Fe_{1-x}Co_x)_2As_2$ and many other iron-based superconductors can be traced back to the excellent purity of the best $Ba_{1-x}K_xFe_2As_2$ single crystals. When impurity scattering in a superconductor increases, this high-energy anomaly due to the Holstein absorption becomes washed out by the overall strong direct absorption setting in already above the quasiparticle excitation gap 2Δ. We thus argued that spectroscopic ellipsometry is particularly sensitive to the energy scale of the mediating boson in clean materials. We further demonstrated that the far-infrared conductivity of both the hole- and the electron-doped 122 iron arsenides can be well described in the framework of a minimal two-band Eliashberg model reduced from its complete four-band counterpart. This fortunate theoretical simplification is justified by the clustering of the superconducting energy gaps into two groups with two distinct values of the superconducting gap.

We then proceeded to investigate the complex dielectric function of high-quality nearly-stoichiometric $Rb_2Fe_4Se_5$ single crystals with $T_c = 32$ K by broadband spectroscopic ellipsometry and time-domain transmission spectroscopy in the spectral range 1 meV $\leq \hbar\omega \leq$ 6.5 eV at temperatures 4 K $\leq T \leq$ 300 K. This compound was found to simultaneously display a superconducting and a semiconducting optical response. It revealed a direct band-gap of $\approx$0.45 eV determined by a set of spin-controlled interband transitions. Below 100 K we observed in the lowest THz spectral range the development of a clear metallic response characterized by a negative dielectric permittivity ε_1, with the bare (unscreened) $\omega_{pl} \approx 100$ meV at the lowest temperatures in the normal state. At the superconducting transition this metallic response was found to exhibit a signature of the superconducting gap below 8 meV, consistent with other experimental probes. Our findings suggested coexistence of superconductivity and magnetism in this compound as two separate phases.

Disentangling the inherent optical response of constituent phases in a phase-separated material can be a daunting task. First of all, at least some information about one of the phases should be obtained independently. Secondly, precise information about the shape, size, and structure of the phase domains must be obtained lest the analysis within any effective-medium theory suffer from extensive uncertainty. To obtain such detailed information about the phase domains in a single-crystalline antiferromagnetic superconductor $Rb_2Fe_4Se_5$ we studied it using a combination of s-SNOM and LE-μSR. We demonstrated that the antiferromagnetic and superconducting phases in this material segregate in the out-of-plane direction into nanometer-thick layers parallel to the iron-selenide planes, while the characteristic in-plane size of the metallic domains reaches 10 μm. By means of LE-μSR we further showed that in a 30-nm thick surface layer the ordered antiferromagnetic moment is drastically reduced, while the volume fraction of the paramagnetic phase is significantly enhanced over its bulk value. Such self-organization into a quasiregular heterostructure indicates an intimate connection between the modulated superconducting and antiferromagnetic phases. Whereas most iron-based superconductors to have exhibited phase separation at early stages of crystal growth, such as $Ba_{1-x}K_xFe_2As_2$ [6], have now been found to display microscopic phase coexistence in a chemically and electronically homogeneous material [7], in the case of the $A_2Fe_4Se_5 (A = K, Rb, Cs)$ compounds significant experimental evidence suggests that the phase-separation phenomenon is intrinsic and, in fact, is likely to give rise to superconductivity in the first place due to the internal pressure effect of the antiferromagnetic layers on the superconducting domains, which modifies the pnictogen height in the superconducting phase and thus optimizes superconductivity. If correct, this picture would suggest that even if the chemical equivalent of the superconducting phase of $Rb_2Fe_4Se_5$ is isolated (and it might have been) it will remain unnoticed due to the absence of superconductivity under ambient pressure.

References

1. Evtushinsky, D. V. et al. (2012). Strong pairing at iron $3d_{xz,yz}$ orbitals in hole-doped $BaFe_2As_2$. arXiv:1204.2432 (unpublished).
2. Sudayama, T., et al. (2011). Doping-dependent and orbital-dependent band renormalization in $Ba(Fe_{1-x}Co_x)_2As_2$ superconductors. *Journal of the Physical Society of Japan*, *80*, 113707.
3. Shimojima, T., et al. (2010). Orbital-dependent modifications of electronic structure across the magnetostructural transition in $BaFe_2As_2$. *Physical Review Letters*, *104*, 057002.
4. Kontani, H., & Onari, S. (2010). Orbital-fluctuation-mediated superconductivity in iron pnictides: Analysis of the five-orbital Hubbard-Holstein model. *Physical Review Letters*, *104*, 157001.
5. Saito, T., Onari, S., & Kontani, H. (2011). Emergence of fully gapped s_{++}-wave and nodal d-wave states mediated by orbital and spin fluctuations in a ten-orbital model of KFe_2Se_2. *Physical Review B*, *83*, 140512.
6. Park, J. T., et al. (2009). Electronic phase separation in the slightly underdoped iron pnictide superconductor $Ba_{1-x}K_xFe_2As_2$. *Physical Review Letters*, *102*, 117006.
7. Avci, S., et al. (2012). Phase diagram of $Ba_{1-x}K_xFe_2As_2$. *Physical Review B*, *85*, 184507.

About the Author

Aliaksei Charnukha was born on April 22nd, 1985 to the family of Sedach Liudmila Ivanauna, an IT engineer, and Charnukha Siarhei Piatrovich, a physicist. He grew up in the capital of Belarus, Minsk, where he attended a mathematical middle school and a linguistic gymnasium thereafter. He then went on to study physics at the Belarusian State University specializing in theoretical physics from 2002–2007. From there he graduated *summa cum laude* (with highest honors and one of the best average grades in the history of the university) upon defending his diploma thesis in theoretical quantum optics entitled 'Diffusive processes with translational shifts' in June 2007.

Inspired by ones of the most prominent experimental cold-atom physicists of the time, including Nobel laureate Wolfgang Ketterle, at a summer school in Crete in July 2007, the author decided to switch to experimental research work and joined Tilman Pfau's group in a Master's degree program at the University of Stuttgart, Germany, in October 2007, concentrating on Bose-Einstein condensation in dilute gases. Upon successful completion of this course of study almost half a year ahead of the usual term in May 2009 he commenced his doctoral research in the solid-state spectroscopy group of Bernhard Keimer at the Max Planck Institute for Solid-State Research in Stuttgart. Therein he carried out experimental investigation of charge-carrier dynamics in a newly discovered family of superconductors, iron pnictides, by means of spectroscopic ellipsometry, near-field optical microscopy, terahertz time-domain transmission spectroscopy, as well as studies of local magnetism via low-energy muon-spin rotation/relaxation. In December 2012 he defended his doctoral dissertation entitled 'Charge dynamics in 122 iron-based superconductors' and graduated *summa cum laude* (with highest honors) from the University of Stuttgart.

A. Charnukha, *Charge Dynamics in 122 Iron-Based Superconductors*,
Springer Theses, DOI: 10.1007/978-3-319-01192-9,

Already as a doctoral student the author attained significant international visibility by publishing research articles in high-profile journals, attending a number of international conferences, some of which as an invited speaker, reviewing more than twenty articles as a referee of Physical Review Letters and Physical Review B, and initiating several fruitful international collaborations.